Education in the 21st Century

Amanda Berry • Cathy Buntting
Deborah Corrigan • Richard Gunstone
Alister Jones
Editors

Education in the 21st Century

STEM, Creativity and Critical Thinking

Springer

Editors
Amanda Berry
Monash University
Clayton, VIC, Australia

Cathy Buntting
University of Waikato
Hamilton, New Zealand

Deborah Corrigan
Monash University
Clayton, VIC, Australia

Richard Gunstone
Monash University
Clayton, VIC, Australia

Alister Jones
University of Waikato
Hamilton, New Zealand

ISBN 978-3-030-85299-3 ISBN 978-3-030-85300-6 (eBook)
https://doi.org/10.1007/978-3-030-85300-6

This Springer imprint is published by the registered company Springer Nature Switzerland AG
The registered company address is: Gewerbestrasse 11, 6330 Cham, Switzerland

Preface

This is the seventh book in a series initiated by Monash University-King's College London International Centre for Study of Science and Mathematics Curriculum, in partnership with the University of Waikato. The Monash-King's College Centre was established in 2002 with initial support from the Monash University Research Fund (New Areas). The Centre for Science, Mathematics and Technology Education at Monash University and Waikato University's Technology, Environmental, Mathematics and Science (TEMS) Education Research Centre have had a formal partnership agreement since 2003 and have worked cooperatively in many areas.

The first book in the series, *The Re-emergence of Values in Science Education* (D. Corrigan, J. Dillon and R. Gunstone [Eds.], 2007, Rotterdam: Sense), considered the state of science education in the twenty-first century through the lens of values. The book presented a 'big picture' of what science education might be like if values once again became central in science education. At the time, the overwhelming experiences of those who were teaching science were in an environment that had seen the de-emphasis of values fundamentally inherent in both science and science education. There was a disparity between the evolutionary process that science was – and still is – undertaking and that undertaken by science education (and school science education in particular).

In the second book, *The Professional Knowledge Base of Science Teaching* (D. Corrigan, J. Dillon and R. Gunstone [Eds.], 2011, Dordrecht: Springer), our intent was to explore what expert science education knowledge and practices may look like in the then slowly emerging 'bigger picture' of the re-emergence of values, which we saw as a logical step from the first book's exploration of values. We noted in the Foreword to this book that the focus of the book was on 'exploring what expert science education knowledge and practices may look like in the emerging 'bigger picture' of the re-emergence of values'.

In the third book, *Valuing Assessment in Science Education: Pedagogy, Curriculum, Policy* (D. Corrigan, R. Gunstone and A. Jones [Eds.], 2013, Dordrecht: Springer), we took what we considered to be another logical next step in the sequence of foci begun with our exploration of values: assessment. The reality of education is that it is assessment that is almost always the strongest force shaping teacher development and behaviour, the implemented curriculum, student approaches to learning, etc. Consequently, the third book considered the 'big picture' of assessment in science education, from the strategic and policy level to that of classrooms. However, while some classroom case studies were presented, they focused more on teachers than students, and so considered assessment more in terms of what teachers plan and do rather than the impacts of assessment on students.

The fourth book, *The Future in Learning Science: What's in It for the Learner?* (D. Corrigan, C. Buntting, J. Dillon, A. Jones and R. Gunstone [Eds.], 2015, Dordrecht: Springer), considered the learning of science in contemporary education: the forms of science that represent the nature of science in the twenty-first century, the purposes we might adopt for the learning of school science, the forms this learning might better take and how this learning happens. Of particular concern was the need to better engage students with their school science and the need to place the burgeoning range of digital technologies into a more informed context than the narrow and uncritical contexts in which they are too commonly being positioned. Additionally, we sought to represent and value the perspective of the learner as an important overarching theme.

The fifth book, *Navigating the Changing Landscape of Formal and Informal Science Learning Opportunities* (D. Corrigan, C. Buntting, A. Jones and J. Loughran [Eds.], 2018, Cham: Springer), championed research involving learning opportunities that are afforded to learners of science when the focus is on linking the formal and informal science education sectors. We use the metaphor of a 'landscape' to emphasise the range of possible movements within a landscape that is inclusive of formal, informal and free-choice science education opportunities, rather than the not uncommon formal sector assumption that the informal sector should somehow serve the formal, and that free choice is not part of education at all. In addition, the book explored opportunities for informing formal school science education via the perspectives and achievements of the informal and free-choice science education sectors.

Then the sixth book, *Values in Science Education: The Shifting Sands* (D. Corrigan, C. Buntting, A. Fitzgerald & A. Jones [Eds.], 2020, Cham: Springer) returned to an explicit focus on values, more than a decade after the first book in this series. In that first book, it was evident that different cultures have different traditions in relation to the place of values in their school science curriculum and that these traditions were being challenged. In this sixth volume, authors reflected on how values are centrally associated with science and its teaching, as well as the wide range of factors that influence science education. These include sociocultural, philosophical and psychological influences; curriculum; the nature of science; formal and informal education settings; the relationship between science, technology, society and the environment; teaching and learning practices; assessment and evaluation; teacher education; and classroom climates. As suggested by the second half of its title, the book sought to capture the persistent but vulnerable nature of values in the face of forceful influences on the education landscape.

In this seventh book, we focus on two major and increasingly global trends that impact directly on curriculum in the sciences and mathematics, engineering and technology. The first of these trends is, unsurprisingly, 'STEM'. This acronym for Science, Technology, Engineering, Mathematics, is today used continuously by, it seems, all groups with concerns for formal education – from classroom teachers to national politicians. It is easy to forget that this ubiquitous use of STEM did not even seriously begin until the present century. It can also be easy to overlook that STEM is used for more than one purpose. Our concern in this volume is with issues relevant to STEM Education. The second trend is for school curriculum in particular to embrace cross-curriculum goals ('competencies' or 'capabilities'), usually intended to be woven into traditional single-discipline subjects. Two of these competences have often been seen as specific parts of science or mathematics curricula – 'creativity' and 'critical thinking'. These two, with STEM Education, are themes through the chapters of this book as these issues are addressed in a wide range of contexts, from individual classrooms to reconceptualising the purposes of STEM Education.

We used the same approach to the creation of this seventh book as we did with the previous six. In seeking to achieve a cohesive contribution to the literature while enabling the authors to assert their own voices without restrictive briefs from us as editors, we again hosted a 3-day workshop involving all the authors to facilitate a more interactive and formative writing process. A first draft of all chapters was distributed prior to the workshop, enabling intensive discussions of individual chapters and feedback to authors and considerations of the overall structure and cohesion of the volume. Authors then rewrote their contributions in the light of the group's feedback. As with the previous books, the workshop was scheduled around the European Science Education Research Association (ESERA) conference in Bologna and took place at the Monash University Centre in Prato (Italy).

This writing process had previously been used very successfully in the production of two other books in which the editors had variously been involved: P. Fensham, R. Gunstone & R. White (Eds.), 1994, *The Content of Science: A Constructivist Approach to Its Teaching and Learning*, and R. Millar, J. Leach & J. Osborne (Eds.), 2000, *Improving Science Education: The Contribution of Research*. More recently, the approach has been adopted by other science education researchers. We believe, strongly, that this process significantly improves the quality of the final product and provides an opportunity for what is sadly a very rare form of professional development for academic researchers – formative, highly collaborative (and totally open) discussions of one's work by one's peers.

We gratefully acknowledge the funding of the workshop through contributions from Monash University and Waikato University and the commitment, openness and sharing of all participants in the workshop.

Clayton, VIC, Australia — Amanda Berry
Hamilton, New Zealand — Cathy Buntting
Clayton, VIC, Australia — Deborah Corrigan
Clayton, VIC, Australia — Richard Gunstone
Hamilton, New Zealand — Alister Jones
October 2020

Contents

1 **STEM Education Matters** . . . 1
Cathy Buntting, Richard Gunstone, Amanda Berry, Deborah Corrigan, and Alister Jones

2 **Creativity and Critical Thinking** . . . 9
Peter Ellerton and Robert Kelly

3 **Fostering Students' Creativity and Critical Thinking in Science Education** . . . 29
Stéphan Vincent-Lancrin

4 **Exploring STEM Learning in Primary Classrooms: In Support of Social Justice Agendas** . . . 49
Bronwen Cowie and Paula Mildenhall

5 **There's Something About James** . . . 63
Cathy Buntting and Alister Jones

6 **STEM, Creativity and Critical Thinking: How Do Teachers Address Multiple Learning Demands?** . . . 81
Deborah Corrigan, Debra Panizzon, and Kathy Smith

7 **Stimulating Creativity and Critical Thinking in Integrated STEM Education: The Contribution of Out-of-School Activities** . . . 99
Léonie J. Rennie

8 **Critical Thinking Across Disciplines in University General Education: Obesity as a Socioscientific Issue** . . . 119
Maurice M. W. Cheng and Jessica S. C. Leung

9 **When Failure Means Success: Accounts of the Role of Failure in the Development of New Knowledge in the STEM Disciplines** . . . 137
Jennifer Mansfield and Richard Gunstone

10 Humanistic Goals for Science Education: STEM as an Opportunity to Reconsider Goals for Education 159
Michael Tan

11 Final Commentary: "Education in the 21st Century: STEM, Creativity and Critical Thinking" 177
Amanda Berry

Editors and Contributors

About the Editors

Amanda Berry is Professor of STEM education in the Faculty of Education at Monash University, Australia. Amanda's research and teaching focuses on the development of STEM teachers' knowledge and the ways in which that knowledge is shaped and articulated through teacher preparation, beginning teaching and in-service learning. Amanda has a strong international profile in the field of science teacher education and science teachers' pedagogical content knowledge (PCK). She has been involved in many research projects focusing on innovations designed to address the quality of teacher professional learning and enhance science teaching and learning in schools and universities.

Cathy Buntting is Director of the Wilf Malcolm Institute of Educational Research at the University of Waikato, New Zealand. With a master's degree in biochemistry and a PhD in science education, her research interests span science education, technology education and STEM education. She is particularly interested in the ways in which digital technology can support learning in these areas, and is director of the New Zealand Science Learning Hub, a vast online resource linking STEM practitioners with schoolteachers and students.

Maurice M. W. Cheng is Associate Professor in the Division of Education, University of Waikato, New Zealand. He is interested in STEM education generally, as well as science and chemistry education. His major research interests include the role of visual representations in science learning and how drawing serves as a tool for learning and in research. He also works in international comparative studies such as TIMSS and PISA.

Deborah Corrigan is Professor of science education in the Faculty of Education at Monash University, Australia. After working as a chemistry and biology teacher, she has worked at Monash University in chemistry and science education, particularly in teacher preparation. Her research interests include industry and technology links, science education, curriculum design, science and STEM education policy, and the values that underpin science education. However, her main research interest remains improving the quality of chemistry and science education so that it is relevant to students.

Bronwen Cowie is Associate Dean of research, Division of Education, the University of Waikato, New Zealand. Her research interests are in assessment for learning, student voice, curriculum development and implementation, and culturally responsive pedagogy and assessment in science and technology education. She has completed a number of large national research projects as well as in-depth classroom studies where she has worked collaboratively with teachers and students.

Peter Ellerton is Curriculum Director of the University of Queensland's Critical Thinking Project and Senior Lecturer in the School of Historical and Philosophical Inquiry. He has a master's degree in science and a PhD in philosophy, and taught in high schools for many years. Peter's areas of research interest include public reasoning, science communication, argumentation and critical thinking in education. His passion is working with educators to enable a teaching for thinking focus across all year levels and subject areas.

Richard Gunstone is Emeritus Professor of Science and Technology Education at Monash University. His published research has involved the areas of learning, teaching and teacher development, curriculum and assessment, at all levels of education; his past work includes editorship of the Encyclopedia of Science Education (Springer, 2015). His current interests include the changing natures of the disciplines of science and of technology in the twenty-first century, and the implications of these changes for school STEM in general and the sciences in particular. In 2014, he received the Distinguished Contributions to Research Award from the National Association for Research in Science Teaching.

Alister Jones is Research Professor and Senior Deputy Vice Chancellor at the University of Waikato, New Zealand. He had a leading role in developing New Zealand's technology education curriculum; has been consulted on educational development in Australia, the UK, USA, Hong Kong, China, Chile and Thailand; and is the director of a number of educational companies. He is a founding board member of the Forum for World Education.

Robert Kelly is Associate Professor in the Faculty of Arts at the University of Calgary, Canada, and Academic Coordinator of the Collaborative Creativity for Social Innovation and Human-Centred Design graduate program at the University's Werklund School of Education. He has pioneered research in the areas of collaborative creativity, creative development, and design thinking for innovation in teacher education through course and program design at his home institution and as an invited visiting scholar at numerous postsecondary institutions in Canada.

Jessica S. C. Leung is Assistant Professor in the Faculty of Education, The University of Hong Kong. Her research focuses on critical engagement with socio-scientific issues and students' understanding of the nature of science. She teaches courses with a focus on socio-scientific issues for pre-service teachers and for students from all faculties at the university.

Jennifer Mansfield is Lecturer in primary and secondary science/biology education in the Faculty of Education at Monash University, Australia. Jennifer previously worked as a scientist and secondary school science teacher, which has led to a keen interest in enhancing the quality of science education. Her research includes exploring ways of ways of articulating the complex nature of teachers' professional knowledge development, supporting teacher thinking about the effectiveness of practical work in science, and teaching science as a human endeavour as a way of enhancing scientific literacy, student interest and learning in science.

Paula Mildenhall is Associate Professor and the Associate Dean of Teaching and Learning in the School of Education at Edith Cowan University in Perth, Western Australia. Over the past 25 years, Paula has held teaching and leadership roles across the education sector in primary and tertiary settings, both in Australia and internationally. Her research focuses on mathematics and STEM education in primary schools and also on how pre-service teachers may effectively learn in tertiary settings.

Debra Panizzon is Adjunct Associate Professor in the Faculty of Education at Monash University, Australia, and Research Analyst for the Teachers Registration Board of South Australia. She has held key positions at the University of South Australia, Flinders University and the University of New England in Armidale, New South Wales. Her research interests include STEM policy, assessment in science and mathematics education, cognition, and critical and creative thinking. Much of her work has involved collaborative research to inform policy in science and STEM education with key educational stakeholders in Australia and internationally.

Léonie J. Rennie is Emeritus Professor in science and technology education at Curtin University and Adjunct Professor at Monash University. Her research focuses on learning in science and technology, particularly outside of school. In 2009, Léonie received the Distinguished Contributions to Science Education through Research Award from the National Association for Research in Science Teaching.

Kathy Smith is Senior Lecturer in the Faculty of Education at Monash University, Australia, with specialisations in primary science education and teacher professional learning. She is also the Course Leader for the Graduate Certificate in STEM Education at Monash University. Kathy began her career as a primary classroom teacher and has been involved in science education in a range of roles for the last 30 years. Her research interests include understanding more about the conditions that promote quality teacher professional learning, primary science education, STEM education, and the relationship between teacher thinking and pedagogy.

Michael Tan is Lecturer (Research Scientists) at the National Institute of Education, Nanyang Technological University, Singapore. His research interests are in curriculum studies and teacher development, with a specific focus on science education. In recent years, he has been studying the implementation of STEM in makerspaces, with an intention towards nurturing students' critical perspective, creativity and wisdom.

Stéphan Vincent-Lancrin is Senior Analyst and Deputy Head of Division at the Organisation for Economic Co-operation and Development (OECD), Directorate for Education and Skills. He leads work on innovation in education and education for innovation. He works on digitalisation in education and has edited the OECD's *Digital Education Outlook 2021: Pushing the Frontiers with AI, Blockchain and Robots*. He is also interested in non-technological innovation, for example, how teachers can foster and assess students' creativity and critical thinking, and more broadly, in the determinants of innovation-friendly ecosystems education.

Contributors

Amanda Berry Faculty of Education, Monash University, Clayton, VIC, Australia

Cathy Buntting University of Waikato, Hamilton, New Zealand

Maurice M. W. Cheng Te Kura Toi Tangata School of Education, University of Waikato, Hamilton, New Zealand

Deborah Corrigan Faculty of Education, Monash University, Clayton, VIC, Australia

Bronwen Cowie School of Education, University of Waikato, Hamilton, New Zealand

Peter Ellerton Faculty of Humanities and Social Sciences, University of Queensland, Brisbane, QLD, Australia

Richard Gunstone Faculty of Education, Monash University, Clayton, VIC, Australia

Alister Jones University of Waikato, Hamilton, New Zealand

Robert Kelly Faculty of Arts, University of Calgary, Calgary, Canada

Jessica S. C. Leung Faculty of Education, The University of Hong Kong, Pokfulam, Hong Kong

Jennifer Mansfield Faculty of Education, Monash University, Clayton, VIC, Australia

Paula Mildenhall School of Education, Edith Cowan University, Perth, WA, Australia

Debra Panizzon Faculty of Education, Monash University, Clayton, VIC, Australia

Léonie J. Rennie School of Education, Curtin University, Perth, Western Australia

Kathy Smith Faculty of Education, Monash University, Clayton, VIC, Australia

Michael Tan National Institute of Education, Nanyang Technological University, Singapore, Singapore

Stéphan Vincent-Lancrin Directorate for Education and Skills, Organisation for Economic Co-operation and Development, Paris, France

Chapter 1
STEM Education Matters

Cathy Buntting, Richard Gunstone, Amanda Berry, Deborah Corrigan, and Alister Jones

1.1 The Pandemic of 2020

Writing this chapter in the latter part of 2020, we are mindful of the many ways that the world has been dramatically defined and transformed by the impacts of COVID-19. The global pandemic has shone the spotlight on a plethora of human concerns that, while not new, have been highlighted in new ways: our interconnectedness and vulnerabilities as a species, our propensity for both compassion and selfishness, the inequities that stretch across and within borders, and ultimately our resilience. Across the world stage, we've seen multiple scenarios playing out – the majority unscripted and continuing to evolve as new developments emerge. Evidence from across the STEM disciplines – science, technology, engineering and mathematics – has shaped individual, community and national responses, but in different ways. While some jurisdictions committed to strategies of reduction and elimination, others pinned hopes on 'herd immunity' and the development of vaccines, and in some places politics superseded acceptance of the spectrum of evidence from the fields of STEM about COVID-19 and how best to respond. In many instances, poverty that preceded the pandemic tragically eliminated choice. In other instances, wilful ignorance had deathly consequences.

C. Buntting (✉) · A. Jones
University of Waikato, Hamilton, New Zealand
e-mail: cathy.buntting@waikato.ac.nz; alister.jones@waikato.ac.nz

R. Gunstone
Monash University, Clayton, VIC, Australia
e-mail: richard.gunstone@monash.edu

A. Berry · D. Corrigan
Faculty of Education, Monash University, Clayton, VIC, Australia
e-mail: amanda.berry@monash.edu; debbie.corrigan@monash.edu

A. Berry et al. (eds.), *Education in the 21st Century*,
https://doi.org/10.1007/978-3-030-85300-6_1

Within these world-wide times of change, the central themes of this book – contemporary education, STEM, criticality and creativity – are even more powerfully relevant. While 'science' may be leading the way in terms of seeking to understand the novel SARS-CoV-2 virus and its variants, epidemiological modelling is based on integrating mathematical, computing and scientific data, processes and interpretations. Innovations – from COVID testing and vaccine development and its mass production and distribution, to digital mechanisms that support contact tracing and the setting up of isolation protocols and facilities – all rely on systems-approaches that integrate all the STEM epistemologies, including critical and creative approaches to knowledge development and deployment. STEM, criticality and creativity remain key to the recovery and rebuilding that societies and economies will urgently need moving forward.

As educators and education researchers, we've been heartened by the incredible innovation and commitment shown by countless teachers across formal and informal education contexts and across all ages of learners. However, inequities have also been exacerbated – some children and families are more exposed to the negative economic and social impacts of regional and national 'lockdowns'. In some contexts children are no longer accessing education at all because they have been pushed prematurely into employment; those children who do not have the digital resources necessary to access 'learning from home' initiatives in contexts where schools have been closed have in many places essentially been excluded from schooling; and school districts and jurisdictions that do not have the resources to support teachers to effectively deliver online learning opportunities have been disproportionately impacted.

How does education research speak with relevance into a climate such as this one? While there was no thought of anything like COVID-19 when we first began working on this book project in 2018, global issues such as sustainable economic development, climate change and disease management were at the forefront of many STEM education initiatives. There has been long-standing recognition in many countries that traditional school structures need to be better preparing young people as citizens of and contributors to a knowledge economy and a twenty-first century society. Our hope in bringing this particular group of academics together to write, discuss, and rewrite was that we would be able to identify key themes that could help to progress the development of STEM education internationally.

1.2 The Power of the 'STEM' Acronym

The rapid popularisation of STEM – science, technology, engineering and mathematics – came after its use in 2001 by staff at the United States National Science Foundation (NSF) (e.g., Hallinen, 2015), primarily when NSF staff were testifying to Congressional committees in Washington. NSF had

> previously used the acronym SMET when referring to the *career fields in those disciplines or a curriculum that integrated knowledge and skills from those fields*. In 2001, however, American biologist Judith Ramaley, then assistant director of education and human resources at NSF, rearranged the words to form the STEM acronym. (Hallinen, 2015; emphasis added)

However, 'STEM' had been used as an acronym at least as early as the 1990s, and often in the context of 'STEM education', including in a 1998 five-year, multi-million dollar grant funded by the NSF, titled "The Science, Technology, Engineering, and Mathematics Teacher Education Collaborative (STEMTEC)" and managed by the already established "STEM Education Institute" at the University of Massachusetts, Amherst – an Institute that continues today.

Over the last two decades the two broad meanings explicitly associated with STEM by the NSF in 2001 (career fields and integrated curriculum) have both been pursued, even though the two are quite disparate. Additionally, they are at times joined by a third meaning, a collective noun for the separate disciplines of the sciences, mathematics, engineering and technology. Quite why STEM has been so pervasive in political rhetoric, public discussion, and education policy when earlier movements such as 'science and technology' clearly were much less influential is likely a consequence of both the short, word-form of the label and the context in which it emerged (a context in which, for example, technologies were very rapidly becoming even more central to daily life while at the same time, and just as rapidly, becoming more and more complex). Whatever the reasons, STEM education has captured global interest and become an educational phenomenon.

Although STEM education is often cited in political and policy domains as being important, even critical, for economic growth, we do not see this as a significant – or helpful – driver for school STEM education. Rather, we see the power of school STEM education in relation to the future of learners more in terms of their "belongingness in society" than in attempting to forecast specific STEM employment possibilities. In general terms, there is an obvious disconnect between arguments that school STEM education should focus on specific preparation for future employment and the common observation that a large proportion of future jobs cannot yet even be imagined. Further, critics of school STEM education for solely economic/employment reasons point to the vast array of data showing there is an ongoing, substantial and still little-changing lack of diversity in STEM career paths and STEM-related employment across gender, ethnicities, cultures and socio-racial groups (e.g., Allen-Ramdial & Campbell, 2014; Estrada-Hollenbeck et al., 2011; Leigh et al., 2020; Pew Research Centre, 2018; UNESCO, 2017).

Our collective view that it is the integrated curriculum meaning of STEM on which school education should focus, and not the career/employment view, is reflected in the chapters of this book. Specifically, we position such integrated STEM education as working to remove barriers between the four disciplinary areas of science, technology, engineering and mathematics, while retaining the value of disciplinary knowledge and skills. Our premise is that everyone needs to be STEM-literate if our communities are to effectively respond to multi-faceted economic, social and environmental challenges such as those foregrounded by COVID-19 and climate change (Corrigan, 2020).

1.3 Cross-Curriculum Capabilities: A Major Curriculum Trend in the 21st Century

As we began our conversations in 2018 about important possibilities for a book to explore contemporary issues with STEM education, we were conscious of global changes in broader curriculum thinking and planning. This included a major shift towards curricula advocating for cross-curriculum 'competencies' or 'general capabilities', with these often generically grouped together as '21st Century skills'. These capabilities are relevant to all subject domains of a curriculum, and are argued to be too significant to the development of learners to be ignored. Such cross-curriculum competencies are now required in many systemic curriculum prescriptions, and are central themes in a number of multi-country projects promoting curriculum thinking (e.g., the 'competencies learners need to succeed' in Fadel, Bialik & Trilling, 2015 and the OECD's *Future of Education and Skills 2030*, n.d.). Among the competencies variously identified in these curricula are two that have long been associated with education in the separate disciplines of STEM – critical thinking and creativity. Current movements in STEM education and broader cross-curriculum competencies both reinvigorate and reinforce the deep importance of creativity and critical thinking in the separate disciplines of STEM and in integrated STEM education. As a consequence, in our invitations to contributing authors for the writing workshop and this subsequent book we specifically identified these two competencies as being of particular interest. (An outline of the invitation and the nature and purpose of the workshop is given in the Preface to this book.)

1.4 The Chapters of this Book

As we note at the beginning of this chapter, the onset of the COVID-19 pandemic came well after the planning of this book and the writing of draft chapters. Nevertheless, and completely unexpectedly, this pandemic provides dramatic global and life-determining examples of science, of mathematics, and of technological advancement. Both creativity and critical thinking have been made clearly evident as central to medicine, social policy, and the reporting of science, mathematics, engineering and technology in mass and social media. When presented with situations where accepted norms are no longer appropriate, available or encouraged, as in the pandemic self-isolation scenarios experienced across the globe, the need for new ways to live and learn becomes fundamentally important. One important lesson from the COVID-19 pandemic has been the demonstration of the profound importance of collaboration (not competition) between professionals of different disciplines – including the STEM disciplines.

Throughout the chapters of this book there is specific concern with the roles of creativity and critical thinking in contemporary and future STEM education. Therefore, Chap. 2 focuses directly on creativity and critical thinking – what these are, and what characteristics they have. One of the authors, Robert Kelly, has a substantial history in researching creativity and teaching about the nature and development of creativity; the other, Peter Ellerton, has the same in critical thinking. The two have co-authored a single chapter because, as they write,

> [t]he application of the concepts of creativity and critical thinking into educational practice across the STEM disciplines [...] requires an integrative approach as these two concepts are so heavily interrelated in practice on so many levels; they are mutually dependent concepts.

In the first substantive section of Chap. 2 Robert Kelly defines creativity in terms of having characteristics of a sequence of thought, a sequence of actions, and a novel adaptive production that occurs within a social context. But, he notes, definition is not enough. It is important for educators, in STEM or otherwise, to not only engage in creativity but to also educate others in creative practice – that is, to be involved in the operationalisation of creativity. His proposed mode of creative development with which educators need to engage includes consideration of the development of collaborative and communication capacity as groups come together to engage in the creative process. This includes ideation and prototyping in the production of a novel and useful solution or artefact. Engaging in the creative process also requires some degree of self-initiated development fuelled by intrinsic motivation, a growing appreciation for the complexity of the disciplines within which one operates, and a strong sense of the discipline expertise one has to contribute to the process.

Peter Ellerton then continues Chap. 2 by addressing the nature and characteristics of critical thinking. He notes that while critical thinking does not have a specific and unique disciplinary home, it has a logical academic home in philosophy because philosophy "provides a rigorous normative framework for understanding critical thinking". Critical thinking is widely accepted to involve skills (e.g., argument construction, evaluation, communication), inquiry values (e.g., values associated with the process of inquiry such as accuracy, reproducibility, coherence), and inquiry virtues (characteristics of an individual critical thinker rather than a process of critical thinking, e.g., open-mindedness, tolerance, honesty, charity). Kelly and Ellerton argue that creativity and critical thinking, in their mutual dependence, are both developed under conditions that include doubt, collaborative investigation, and shared commitment to completion of the task or goal that initiated the development. None of the entities at the heart of this book – STEM education, creativity and critical thinking – exist in isolation. They are entwined. Understanding how they are entwined and in what contexts, and how creativity and critical thinking are central to STEM education, is a consistent focus throughout the subsequent chapters of the book.

In Chap. 3, Stéphan Vincent-Lancrin considers what the capabilities of creativity and critical thinking might specifically involve when exemplified in school science education, and how one might consider the development of each of these among students. He describes a multi-country OECD project to develop both domain-general and science domain-specific conceptual rubrics for creativity and for critical thinking, and lays out the products of the project. He then illustrates the science-specific rubric via application in two different science curriculum units, and outlines teaching and learning strategies aligned to both the development of creativity and critical thinking, and to the rubrics. One of the powerful outcomes from using both domain-general and domain-specific conceptual rubrics is in them providing a shared language in which to talk about creativity and critical thinking. Vincent-Lancrin next uses science as an example of how such concepts can be developed within a specific discipline with this shared language. Readers are invited to extrapolate from these insights to consider the implications for such conceptual development not only in the discrete disciplines within STEM but also within integrated STEM education approaches in their own contexts.

While creativity and critical thinking continue to be emphasised in Chap. 4, Bronwen Cowie and Paula Mildenhall take as central themes notions of social justice, equity, and the important role of empathy in engaging learners in authentic STEM contexts, including genuine considerations of possible action. Using three examples, Cowie and Mildenhall convincingly demonstrate that it is necessary to walk in the shoes of others if students are to give authentic, respectful consideration to taking some agency over potential subsequent actions in response to a real STEM issue. Specifically, the three different vignettes from STEM primary school classrooms demonstrate the differing paths empathetic actions can take for the learner. They also demonstrate that while knowledge is necessary, it is not sufficient for students to be willing and able to take constructive action. It is the development and exercising of empathy, alongside critical and creative thinking, that assists such action.

The use of case studies of primary school classrooms to illustrate significant issues in STEM education and creativity and critical thinking continues in Chapters 5 and 6. First, Cathy Buntting and Alister Jones take us into a senior primary school STEM classroom in which an experienced, committed and clearly expert STEM teacher has his class building simple hydraulic machines. In this detailed case study, Buntting and Jones demonstrate the nature of student learning in this specific integrated STEM context, and, particularly, the importance of focussed conversations between teacher and students in supporting the development of the students' creative and critical thinking. The chapter is a powerful illustration of the ways in which the multiple knowledges held and used by expert teachers have major impact on the development of quality student learning in their classrooms.

In Chap. 6, Deborah Corrigan, Debra Panizzon and Kathy Smith provide two case studies of individual teacher development concerned with implementing STEM teaching with a focus on creativity and critical thinking. Corrigan, Panizzon and Smith begin by providing the background to their own thinking about STEM, creativity and critical thinking before laying out the two examples of teachers who, after experiencing a relevant extended professional learning programme, consider the implementation in their classroom of STEM with a focus on developing creativity and critical thinking. The chapter identifies some of the decisions that influenced the strategies, practices and approaches used by the participating teachers in this implementation.

In Chap. 7, Léonie Rennie takes us outside the classroom to explore the effectiveness of integrated curricula with an out-of-school component in encouraging students to develop their STEM understanding and skills by developing their school disciplinary knowledge in authentic, real-world contexts. Using three quite different examples of students working on issues that are important to their local community, she considers matters relating to social values and diversity that are involved in giving opportunities to students to develop their sense of social and ecojustice (see also Chap. 4). Rennie also links explicitly with Vincent-Lancrin's work (see Chap. 3) by analysing the students' STEM learning in terms of the OECD-developed dimensions of creativity and critical thinking – inquiring, imagining, doing, and reflecting.

In Chap. 8, Maurice Cheng and Jessica Leung shift the focus of the learning to higher education, presenting a case study of undergraduate students engaging with obesity as a socioscientific issue, or SSI. As such, this chapter considers the similarities and differences between integration of disciplines through STEM, and through an SSI – while both can be characterised as approaching integration via a specific context, issue or problem, the SSI integration has an overriding commitment to the centrality of social context. Thus, this chapter presents a different perspective on the possibilities for interdisciplinary learning beyond (but related to) STEM. There is a clear focus on critical thinking throughout the unit, as well as on the prevalence of, and students' adherence to, particular ways of thinking and perceived ways of thinking associated with disciplines such as science. When emphasis in the unit is placed on challenging these ways of thinking, some shifts in terms of both adherence to and prevalence of particular ways are shown. The authors draw comparisons between a technocratic dimension of thinking and an emancipatory dimension of thinking.

Chapters 9 and 10 move away from classroom examples as the context for considerations of aspects of STEM, creativity and critical thinking, but in very different ways. In Chap. 9, Jennifer Mansfield and Richard Gunstone consider how 'failure' is represented in the different disciplines represented by STEM. The role and nature of failure in the development of new knowledge in each discipline is described, alongside how each discipline formally represents the role of failure in knowledge development. The ways in which failure contributes or otherwise to the school learning of each of the STEM disciplines is also considered briefly, and, unsurprisingly, shown to be essentially quite separate from the role of failure in the development of new knowledge in the discipline.

Michael Tan, in Chap. 10, explores something very different. He notes that it is currently all-too-common for educators (both within and beyond STEM education) to associate STEM education primarily with engaging lessons in which student learning is motivated by the construction and/or use of interesting devices. He argues that the prominence of STEM education should enable much more than this; it is an opportunity to reconsider some fundamental goals for education. Tan's central thesis is developed through discussions of the current nature of STEM education, problems with contemporary approaches to science and technology education in particular, and humanistic goals for education. The chapter brings these three issues together in a concluding discussion of the question, "What is the humanist opportunity in STEM education?"

Finally, in the concluding chapter, Amanda Berry offers a reflective commentary on the contribution of the book as a whole, taking up the idea supported by the book's contributors that STEM education can and needs to be much more than an educational reform agenda to supply a future workforce. Berry frames her commentary through the ways in which STEM education opens up an important opportunity for teachers as agents of educational innovation, and how the STEM agenda might be utilised as a means to enable and empower teachers to bring about pedagogical transformation within their own educational contexts, and enrich teachers' professional knowledge of practice. Drawing on the book's chapters, Berry proposes some potential pathways to drive the development of a future STEM education agenda that embraces these goals and that reinforces the central importance of creativity and critical thinking in this endeavour.

References

Allen-Ramdial, S.-A., & Campbell, A. (2014). Reimagining the pipeline: Advancing STEM diversity, persistence, and success. *Bioscience, 64*, 612–618.

Corrigan, D. (2020). *Education futures spotlight report 2. Implementing an integrated STEM education in schools: Five key questions answered.* Monash Education Futures: Monash University. Retrieved from https://cog-live.s3-ap-southeast-2.amazonaws.com/n/1271/2020/Aug/11/FJpR1YB8c0JO4Y9mLLht.pdf

Estrada-Hollenbeck, M., Woodcock, A., Hernandez, P. R., & Schultz, P. W. (2011). Toward a model of social influence that explains minority student integration into the scientific community. *Journal of Educational Psychology, 103*, 206–222.

Fadel, C., Bialek, M., & Trilling, B. (2015). *Four-dimensional education: The competencies learners need to succeed.* Boston, MA: The Center for Curriculum Redesign.

Hallinen, J. (2015). STEM. In *Encyclopedia Britannica online*. Retrieved from https://www.britannica.com/topic/STEM-education

Leigh, K., Hellsing, A., Smith, P., Josifovski, N., Johnston, N., & Leggett, P. (2020, July). *Australia's STEM workforce*. Canberra: Office of the Chief Scientist. Retrieved from https://www.chiefscientist.gov.au/news-and-media/2020-australias-stem-workforce-report.

OECD. (n.d.) *OECD future of education and skills 2030 project*. Retrieved from https://www.oecd.org/education/2030-project/about/

Pew Research Centre. (2018, January). *Women and men in STEM often at odds over workplace equity*. Retrieved from https://www.pewsocialtrends.org/2018/01/09/diversity-in-the-stem-workforce-varies-widely-across-jobs/

UNESCO. (2017). *Cracking the code: Girls' and women's education in science, technology, engineering and mathematics (STEM)*. Paris: UNESCO.

Chapter 2
Creativity and Critical Thinking

Peter Ellerton and Robert Kelly

Abstract The twenty-first century has seen a rapid growth of curriculum initiatives that consider the development of cross-curriculum competencies as a core issue, and significant for every discipline area. Both because of such cross-curriculum developments and because of the nature of STEM itself, the integration of the particular core competencies of 'creativity' and 'critical thinking' across the STEM disciplines has also grown rapidly in educational importance. Creativity and critical thinking in education are best viewed from the perspectives of both learner development and teacher expertise, with the attributes specific to each concept appropriately seen as increasing in sophistication or complexity over time. A broad examination of each of the two concepts and their interrelatedness, and the consequent implications for educational practice concerned with developing them, creates a lens through which to view the application of creativity and critical thinking across the complexity and diversity of the STEM disciplines and their integrated forms.

Keywords Creativity · Critical thinking · STEM · Learners · Curriculum · Pedagogy

2.1 Introduction

In defining and describing twenty-first century competencies in educational practice it is increasingly common to see creativity, critical thinking, collaboration, and communication in various configurations as 'core curricular competencies', that is

P. Ellerton
Faculty of Humanities and Social Sciences, University of Queensland, Brisbane, QLD, Australia
e-mail: peter.ellerton@uq.edu.au

R. Kelly (✉)
Faculty of Arts, University of Calgary, Calgary, Canada
e-mail: rkelly@ucalgary.ca

A. Berry et al. (eds.), *Education in the 21st Century*,
https://doi.org/10.1007/978-3-030-85300-6_2

as competencies whose development is significant in all the discipline areas of the curriculum. The Organisation of Economic Cooperation and Development (2018) initiatives *Teaching, Assessing and Learning Creative and Critical Thinking Skills in Education* (OECD, 2018a) and *Fostering and Assessing Student's Critical and Creative Thinking Skills in Higher Education* (OECD, 2018b) are major exemplars, and testimony to a global dynamic to bring creativity and critical thinking in particular into the milieu of educational practice. The *Partnership for 21st Century Learning (P21)* (2018) in the United States and its 21 leadership states, is committed to the implementation of P21's Framework for 21st Century Learning. This Framework encompasses the four c's of creativity, critical thinking, collaboration and communication.

Both as part of these cross-curriculum developments and because of the nature of STEM *per* se, the integration of creativity and critical thinking into educational practice across the STEM discipline spectrum has rapidly grown in educational importance this century. There are many new and emerging regional, national and international curricular initiatives recognising these concepts in practice as educational imperatives, some of which are alluded to in other chapters in this volume. The diversity and complexity of the STEM disciplines requires an integrative educational approach that seeks to foster seamless interdisciplinarity and transdisciplinarity; this in turn facilitates complex creative problem solving and resultant innovative research and production. The application of the concepts of creativity and critical thinking into educational practice across the STEM disciplines also requires an integrative approach as these two concepts are so heavily interrelated in practice on so many levels; they are mutually dependent concepts. Critical thinking permeates every aspect of creative practice and creative development. Creative practice is catalytic in the acquisition and growth of the skills, dispositions, habits, values and virtues central to growth in complexity of critical thinking. The combination of the concepts of creativity and critical thinking in educational practice provides a very potent avenue for integrative STEM educational practice.

Creativity and critical thinking in educational practice are best viewed from a learner growth and development perspective where the acquisition of attributes specific to each concept are viewed as increasing in complexity over time. A detailed examination of each concept, their interrelatedness and the implications for educational practice creates a lens through which to view the application of creativity and critical thinking across the complexity and diversity of the STEM disciplines.

2.2 Creativity in STEM Education

For decades traditional educational discourses have had major emphasis on knowledge transfer and corresponding assessment of learner retainment of this knowledge, with the separate disciplines of STEM being exemplars of such emphases. When applying the concept of creativity across the STEM disciplines the central educational challenge becomes how to operationalise creativity in educational practice against a backdrop of such traditional educational discourses. These discourses are often characterised as being consumptive-intense, risk-averse and generally passive in character (Waks, 2014). By stark contrast, an educational culture of creativity requires a highly interactive, experiential culture with low risk-aversion that is conducive to collaborative ideation and prototyping over time in any discipline context, and particularly across the STEM disciplines. Creativity in STEM education has the potential to be a highly integrative educational dynamic across the diversity and complexity of the STEM disciplines. The resultant educator and learner virtues are far more conducive to producing the original research and action across integrated STEM disciplines so necessary to, for example, meet United Nations Sustainable Development Goals (2019) and other complex regional and global problems that have yet to emerge.

An educational discourse focused on creativity in STEM education begins with an examination and clarification of the concepts, including related vocabulary associated with the term 'creativity'.

2.2.1 Creativity – Definitions, Vocabulary and Related Concepts

In everyday contexts the word creativity is often used interchangeably with the concepts of originality, innovation, divergent thinking and idea generation. However, no matter how closely associated with the word creativity each of these terms may be, these are distinctly different concepts in their own right. The etymology of the word creativity is rooted in the Latin word *creare* meaning to make or produce in a physical sense (Gotz, 1981; Piirto, 2004). The clear implication here is that for an activity to be regarded as creative the activity has observable physical results. Piirto adds that "to be *creative* is to be *originative*. *Originative* implies to make something new. *To be creative then, is to make something new or novel"* (p. 6). There is an implied process here that creativity involves moving from thought to form. Lubart's (2000) definition of creativity as "a sequence of thoughts and actions that leads to novel adaptive production" (p. 295) provides a foundational definition of the concept of creativity that reflects the core attributes of the term. It is important to keep in mind the three attributes of this definition as it will clarify usage and application in an educational context and preclude confusion with other related terms (Kelly, 2020). The three identifiable attributes are:

- A sequence of thought – imagination and ideation that lead to actions
- A sequence of actions – the making of an observable physical form in the currency or medium of the discipline or fields where the creative production is occurring.
- Novel, adaptive production – the creative outcome displays original or novel qualities.

These attributes are useful in differentiating and contextualising related concepts and vocabulary when discussing and applying the terms creativity and creative process in educational practice.

Wallas (1926), in his work *Art of Thought*, presented one of the first models of the creative process. In this model, the creative process is presented as a four – stage process consisting of preparation, incubation, illumination and verification. Thus the Wallas stage model of the creative process embodies the foundational tenets of moving from thought to tangible form. This early stage model would inform many variations of stage theory of creative process in subsequent decades. In the time that followed, a general transition occurred from the creative process being perceived as largely existing in the realm of the subconscious to a perception where the process could be deliberately enhanced through actions such as ideation that Osborn (1963) would later add. The design process, as articulated by the international design firm IDEO (2012) in association with Stanford University's 'd.school', resonates very strongly with classic stage theories of creativity with their five stages of discovery, interpretation, ideation, experimentation (prototyping) and evolution. In this context, design process can be viewed on one level as a transdisciplinary application of creative processes for contextual creative problem solving.

The U.S. psychologist J. P. Guilford's (1959) benchmark work regarding traits of creative individuals expanded the vocabulary in this field, with many terms closely associated with the concept of creativity sometimes used as synonyms for creativity in general usage. Guilford's traits of creative individuals encompassed the following:

- Fluency of thinking – The ability to think effortlessly, especially in ideation.
- Flexibility in thinking – The capacity to readily abandon old ideas and accept new ones.
- Originality – The capacity to come up with unusual ideas that are remote from previous concepts.
- Redefinition – The capacity to give up old interpretations of concepts or objects and replace them with new ones.
- Elaboration – The capacity to fill in details or to add in details to a general scheme.
- Tolerance of ambiguity – The willingness to accept some uncertainty, precluding rigidity in thinking.
- Convergent thinking – Thinking toward a solution or problem resolution by sorting through possible alternatives.
- Divergent thinking – open-ended thinking and the generation of numerous, potential, problem-resolution alternatives.

All of these terms (fluency, flexibility etc) are closely related to the concept of creativity but none are synonyms for creativity. The terms divergent thinking and originality are often used interchangeably with the word creativity, yet while these refer specifically to particular components of the creative process or characteristics of the creative process, neither can be wholly substituted for the word creativity. Divergent thinking refers to the generation of alternatives that are potential problem resolutions. Divergent thinking is just part of the longitudinal thought-to-form process, albeit an important component, but not the complete process. Originality refers to an attribute of degree of novelty of a creative outcome relevant to a previous outcome.

Plucker et al. (2004) echo the core attributes of Lubart's (2000) definition of creativity, given above, and add the importance of the social context of the product utility and novelty: "the interaction among aptitude, process and environment by which an individual or group produces a perceptible product that is both novel and useful as defined within a social context" (p. 90). The addition of social context speaks to how a resultant product, when one has engaged in a sequence of thoughts and actions, is perceived as creative or not. When applying the concept of creativity within STEM educational practice it is important for educators and learners to know why something qualifies (or does not) as a creative outcome.

2.2.2 *Individualistic and Sociocultural Definition Contexts*

Sawyer (2012) describes two definitional strands around the concept of creativity, one rooted in an individualistic approach and the other in a sociocultural approach. The *individualistic* definitional approach implies that the novel production is relative to the individual's previous production, whether this production is new to the world or not. For a young learner involved in a STEM education initiative this could involve the creation of something new or novel for that learner or learner's group that is already known in the STEM fields but is new or novel relative to the learner's previous creative STEM production. The *social/cultural* definitional approach implies that the novel production is useful and valuable to a field or domain that is valued by a suitably knowledgeable social group. As the creative output of a STEM learner grows in complexity and sophistication it gains relevance to its particular field or domain. A good example of this is an Irish student Fionn Ferreira, an 18-year-old from West Cork, Ireland who developed a methodology to remove microplastics from water (Bowers, 2019). Ferreira used a combination of oil and magnetite powder to create a ferrofluid in the water containing microplastics. The microplastics combined with the ferrofluid, creating a product for which Ferreira used a magnet to remove the product and thus leaving only water. This work was original enough to the field to warrant a substantial award through a Google sponsored competition for 13–18 year-olds. His previous STEM work as a younger learner showed his evolution from the production of original work from an individualistic perspective to a point where his original work grew in sophistication and complexity to have social/cultural relevance and be original in this field.

2.3 Creative Development

Creative development (Fig. 2.1) is seen as the growth from the natural human disposition of intuitive/adaptive creativity to the development of capacities to engage in increasingly more complex, sustained creative practice characterised by original research and production that has greater sociocultural relevance and importance. Sustained original research and production is characterised by imaginative vision that leads to recurrent iterations of idea generation, prototyping and refinement over time. While creative capacity is viewed as the level of complexity in which one can engage in creative practice at a point in time while creative development (see Fig. 2.1) is viewed as the growth in creativity capacity of an individual or organisation over time (Kelly, 2012, 2016, 2020). Design thinking, and design practice are viewed as contextual creative problem solving where creative processes are applied across discipline contexts to solve problems.

Understanding the dynamics of creative development in educational practice is necessary to facilitate assessment of creative growth and development in STEM education or any other educational context.

Creative development (Fig. 2.1) is characterised as the growth in creative capacity of an individual or group through nine concurrent, interrelated developmental lenses within four major developmental groupings. This construct has an affinity with Amabile's (2012) and Amabile and Pratt's (2017) componential theory of creativity where they describe four components that are necessary for any creative response: domain-relevant skills, creativity-relevant processes, and intrinsic task motivation and the social environment in which the individual is working. The creative development strands encompass the following:

Foundational Developmental Strands - *Collaborative development and communicative development* are viewed as co-foundational developmental strands. These strands represent the transition from ego-centric to group-centric dispositions that is conducive to idea generation and prototyping through the development of collaborative innovation networks. These foundational developmental strands are essential for creative growth and development within an integrated framework across the diversity and complexity of the STEM disciplines. The high levels of communicative and collaborative interactivity required for creative production across the diversity of STEM disciplines have educational implications for creative development in both the individual and group contexts.

Collaboration can be characterised by the following attributes relating to the group within which the collaboration occurs:

(a) The group exists for novel, adaptive creative production
(b) The diversity of group members is leveraged towards shared creative production
(c) A shift from ego-centric to group-centric dispositions of individual group members is central to maximising the creative potential of the group
(d) Creativity requires high interactivity communication characterised by open-mindedness, empathy, flexibility, active and deep listening, respect and honesty.

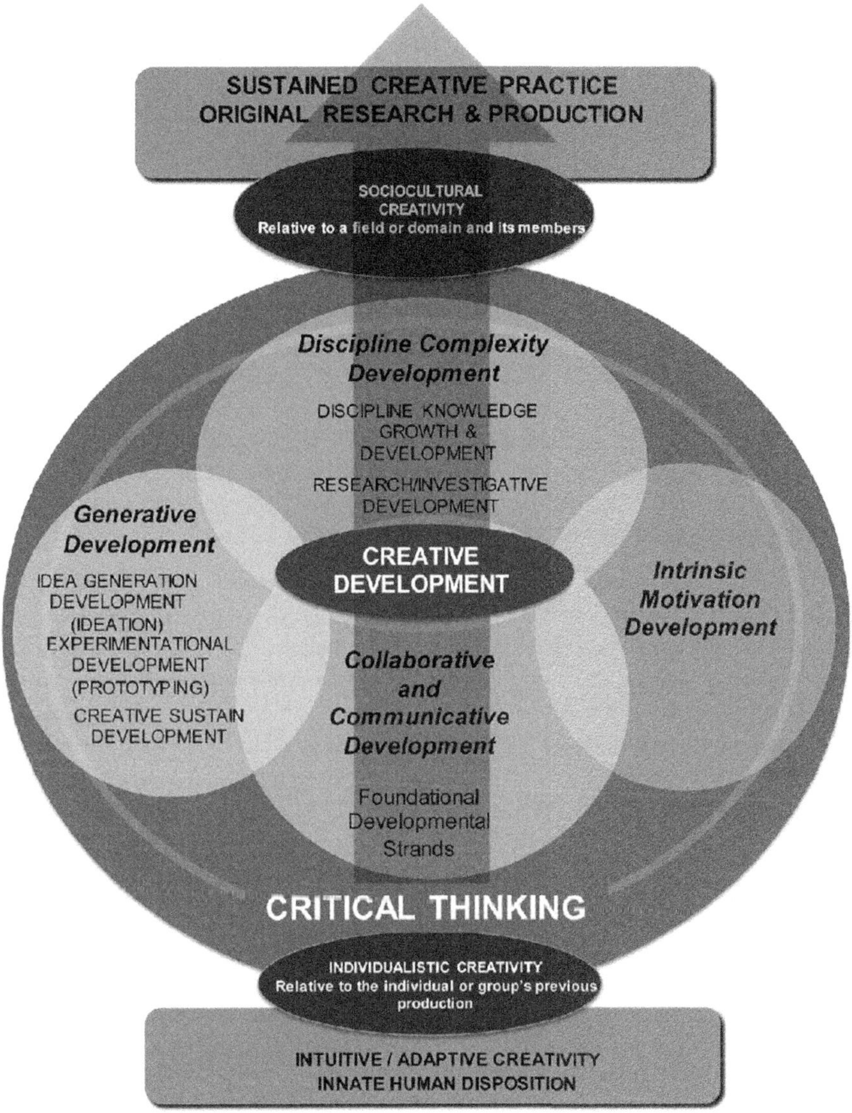

Fig. 2.1 Creative development. (From Kelly, 2016, p. 11)

(e) The potential creative outcomes of the collaborative group would, at best, be highly unlikely to come from an individual.

Communicative development goes hand in glove with collaborative development because of the high interactivity necessary to ideally enable collaborative growth. Foundational collaborative and communicative development are both essential for establishing an educational culture that is conducive to growth in creative capacity through the ensuing developmental strands. Higher interactivity is conducive to higher and more rapid ideation and prototyping development enabling greater creative outcome potentials. Gloor (2017) corroborates this notion through the comprehensive analysis of communication among employees of innovative organizations that is conducive to collaborative creativity.

2. *Intrinsic motivation development* represents inventive growth and development transition from an extrinsic motivational perspective to a disposition of intrinsic motivation where creative development will ultimately lead to learner-initiated innovation and invention. Amabile and Pratt (2017), through their dynamic componential model of creativity, speak to the positive catalytic impact of intrinsic motivation development on creating higher innovation potentials. This implies need for fundamental change to traditional education discourses in STEM disciplines that focus almost exclusively on discipline content transfer and the reward, through assessment, of learner retainment. This, in turn, points to a need to create educational space for learner initiated and owned research, design and invention across learning levels and disciplines.that moves well beyond the consump[tion and retainment of discipline data.
3. *Generative development (Idea Generation Development, Experimentation/ Prototyping Development & "Creative Sustain" Development)* speaks to the growth and development of core creative processes that encompass ideation, experimentation, prototyping and refinement of creative resolutions over time. As a learner develops creative capacity they are able to engage in increasingly longer and more complex iterations of ideation and prototyping resulting in greater "creative sustain". Complex creative initiatives have the potential to involve creative interactivity with regional and global collaborative innovation networks. As learners increase in creative capacity, they are able to engage in increasingly more complex creative initiatives that span months and years.
4. *Discipline complexity development* is the growth and development of the capacity to obtain increasingly more complex knowledge and processes of discipline areas specific to a creative endeavor. This encompasses *Research/ Investigative development* representing a developmental transition from solely passive research disposition to an active, experiential, interactive, investigative disposition. A transition to a disposition of intrinsic motivation enhances the investigative virtue while providing a more meaningful learner context for discipline content acquisition.

These developmental strands are not meant to be viewed in a linear perspective but rather as concurrent, interrelated developmental components that ebb and flow relative to each other depending where one is within a longitudinal, creative initiative. They represent an educational/ecosystem perspective on the interrelatedness of development strands necessary to enable a culture of collaborative creativity across the discipline spectrum and they lend themselves to learning experience design and assessment design for creative development. This facilitates shifts in emphasis from educational cultures focused disproportionately on discipline content acquisition and commensurate assessment regimens to a pedagogical discourse enabling increasingly complex collaborative, creative production by both educator and learner.

2.3.1 Critical Thinking and Creative Development

Critical thinking pervades every aspect of creative practice and creative development. *Critical thinking* includes the effective synthesis and analysis of all gathered and generated information and alternatives to inform decisions leading to problem resolution throughout the creative process. It is the constant evaluation and justification of information and ideas that are absolutely central to advancement of creative production and the growth in creative capacity of the learner. Fig. 2.1, *Creative Development,* shows how critical thinking permeates the educational ecosystem of creative development. It is important to view creativity and critical thinking in concert to fully understand their interrelatedness and the educational ecosystem they enable to maximise educational potentials in STEM education. The examination of the concept of critical thinking in this educational context follows.

2.4 Understanding Critical Thinking

Critical thinking is not necessarily owned, or even best propagated, by any particular discipline. It does, however, find an academic home in philosophy. The reason for this is quite straightforward: questions of what makes for a good reason, why the answer to those questions should compel us to accept such reasons and even the nature of rationality itself are philosophical questions. In other words, philosophy provides a rigorous normative framework for understanding critical thinking. This does not mean that critical thinking can only be developed in philosophy, only that a full study of critical thinking must include a philosophical analysis.

2.4.1 What Is Critical Thinking?

While definitions of critical thinking are many and varied, it is broadly agreed that core technical skills involved with critical thinking include those of argument construction and evaluation (and within these the use of analogy and generalisation), collaborative reasoning skills, and communication skills including effectively linking good thinking and writing. Beyond this technical knowledge, there are also inquiry values, also understood as epistemic values, which are those things we value in the process of inquiry itself. These include accuracy, precision, simplicity, reproducibility, coherence, relevance and a range of other values which enhance inquiry.

The ability to apply these values effectively is essential to critical and empirical inquiry and hence to critical thinking. Inquiry virtues (taking an Aristotelian stance on virtues) are understood as characteristics of individual critical thinkers, rather than characteristics of a process of inquiry. These include open-mindedness, tolerance of other views, intellectual honesty and humility, and the principles of honesty and charity. Driving all this is the imperative to seek new possibilities, reinterpret old problems and question underlying assumptions.

Given this admirable set of attributes, it is natural that the term 'critical thinking' is used frequently and with familiarity in education. It is found in aspirational statements about student outcomes and in the description of any number of tasks that involve a degree of cognitive complexity. That a particular task or curriculum includes elements of critical thinking is an easy and common claim, but common usage does not imply a common understanding of what critical thinking is or how it is to be recognised, directed, planned for or assessed.

The varied definitions of critical thinking together form a clutch of statements each of which is reasonably coherent with the others but seems on its own insufficient to capture the full bloom of cognitive abilities we might associate with a paradigmatic critical thinker. For example, Siegel (1989) speaks of critical thinkers as "appropriately moved by reasons", while Mulnix (2010) says they are capable of recognising and making inferential connections. Paul and Elder (2008) believe critical thinking is about directing one's thinking and willfully subjecting it to standards of evaluation. Willingham (2008) thinks critical thinking includes "seeing both sides of an issue, being open to new evidence that disaffirms your ideas, reasoning dispassionately, [and] demanding that claims be backed by evidence…" (p. 21). One of the most cited definitions in the literature is that of the APA Delphi Report (1990): "purposeful, self-regulatory judgment which results in interpretation, analysis, evaluation, and inference, as well as explanation of the evidential, conceptual, methodological, criteriological, or contextual considerations upon which that judgment is based" (Facione, 1990, p. 3). None of these definitions or statements of necessary attributes seem to contradict the others, but none seem complete, or at least completely descriptive, in themselves.

In some conceptions, critical thinking is strongly tied to expertise and subject area knowledge, so much so that critical thinking without these things seems improbable (see, Willingham, 2019, for example). But this is a difficult case to make, as several people (Mulnix, 2010; Scriven & Paul, 2011; Van Gelder et al., 2004) have pointed out and a range of empirical investigations relating to transferability and continuity of thinking skills (e.g. Topping & Trickey, 2007) have suggested. The American Association of Colleges and Universities (2013) found that 93% of employers believed "a candidate's demonstrated capacity to think critically, communicate clearly, and solve complex problems is *more important* than their undergraduate major" (p. 1, emphasis in original). This statement cannot be reconciled with a conceptualisation of critical thinking developed and used only in discipline knowledge. There is something about thinking well that is transferable, but how can this be expressed?

2.4.2 Skills, Dispositions, Habits of Mind and Inquiry Virtues

To begin, critical thinking must go beyond simply understanding the complexities of problem solving in any particular domain, otherwise any thinking that goes beyond mere recall could, logically, be called 'critical'. While this might be an acceptable definition to some, it is in one sense too broad to deliver a sharp educational focus and in another sense too narrow since it focuses only on a skill set contextualised in the domain.

We noted above that critical thinking embodies a set of skills, values and virtues. Let us now look at these ideas in more detail with attention to their nature and role in critical thinking. Many thinking skills used in discipline-specific contexts are elements of critical thinking, including those of analysis, evaluation, justification and others involved in problem solving and decision-making processes. These skills, generally understood as cognitive skills, are useful in the interpretation, manipulation and creation of knowledge. But it is not enough that we are doing these things in context to claim we are thinking critically—we might be doing them quite poorly or not understand how to apply them well. Most definitions of critical thinking, including those mentioned above, do not restrict themselves to a set of cognitive skills or the context in which they are applied, they also refer to characteristics of the thinker, including habits of mind, thinking dispositions, and virtues. They together contribute to a "critical spirit" (Siegel. 2017) that both animates and guides critical thinking.

A discussion of some terms will be useful here. A habit, generally understood, is a behaviour that has become automatic through frequent expression. But this general definition does not capture the richness of how the term habit is used educationally, particularly in the context of a habit of *mind*. Habits of mind are considered as moving beyond simple reflexive, habitual behaviour and include thinking flexibly, questioning and posing problems, and thinking interdependently (Kallick & Costa, 2008). Being concerned with thinking rather than simply reacting, habits of mind invoke some degree of attention and curation of thought and hence act as the motivation for intelligent action (Dewey, 1938; Dottin, 2009).

Dispositions and habits of mind have a complex relationship in the literature. Siegel (1989), for example, defines a thinking disposition as "a tendency, propensity, or inclination to think in certain ways under certain circumstances" (p. 209), but says little regarding habit. Thornton (2006) offers that (thinking) dispositions themselves are habits of mind, making the former a sub-category of the latter, or perhaps drawing an equivalence between the two. However we might understand this relationship, it does seem clear that we are talking about something more than a basic skill. Siegel (2017) also observes that dispositions can be apparent through a variety of manifestations. A developed disposition to think collaboratively, for example, could be apparent in the sharing of draft proposals with colleagues, actively seeking the advice of others, testing ideas in group settings or choosing to work in the presence of colleagues rather than individually. These are different behaviours, but they point to a particular disposition. Altan, Lane and Dottin (2017) show this is also true for habits of mind. For the purposes of this work let me consider them as, if not synonymous, then at least inseparable. There is an important significant pedagogical point to make here. Thinking dispositions and habits of mind have an important role in establishing how a student thinks and exist separately from any context application. Because they can manifest in a variety of ways, they cannot be equated with behaviours. In terms of developing them, therefore, they cannot be arrived at solely through repetitive acts or learned through unquestioning submission to particular tasks.

Bailin and Battersby (2016) call for an analysis of critical thinking through a virtues framework, arguing that critical thinkers also possess a range of inquiry virtues described by their dispositions, habits and, importantly, a concern for the quality of reasoning in which they engage. These virtues include the oft cited open-mindedness, understanding that knowledge is fallible and the implications of this for inquiry, a willingness to change one's mind, a dedication to reasoned argument above assertion and so on. It may appear that we are again discussing dispositions, but there is an important distinction between dispositions and virtues. As Annas (1995) notes, virtue suggests "some kind of intellectual structure, accessible to the reflective agent" (p. 233). A move to virtues takes us beyond tendencies and habits into a willful, intellectual and value-laden approach to thinking.

> The aspect that is captured in the notion of virtue that is missing in the notion of disposition is that of valuing or appreciating. A virtue is not just a tendency to behave in a certain way but a tendency to do so based on an appreciation or valuing of the enterprise. (Bailin & Battersby, 2016, p. 368)

This care, this understanding of the value of a disposition or habit and why it is important in inquiry, is, according to a virtues account of critical thinking, what separates the virtuous inquirer from those who have simply developed dispositions or habits of mind.

The characteristic that separates effective critical thinkers from the rest of us, including subject area experts with highly specialised and well-developed discipline knowledge, is the ability to evaluate the quality of their own thinking, and that of others, using norms and standards of good reasoning that are developed collaboratively in the thinking classroom. They can also explain how these norms and standards, such as the appropriate application of inquiry values, are derived and their role in good (effective) inquiry, including in discipline methodology. As much as the type of thinking, therefore, it is the quality of thinking that matters; and, to go one step further to virtue, to understand why it matters.

2.4.3 Developing Critical Thinking

It seems clear that critical thinking is experiential, that is, it is something that you *do*. Like all things that you do, there is only so much you can be told about it. Ultimately, you need to practice it yourself and, in the process, get feedback on how to improve. It is a knowing *how* rather than a knowing *that*. An analogy is learning to surf. With all the surfing knowledge in the world available for you to read or view (the knowledge *that*) it would not be sufficient to become a good, or even competent, surfer (the knowledge *how*). At some stage you would need to get on a surfboard and attempt to surf. Feedback on the quality of your surfing would then be necessary to improve, even if that feedback only comes from your own sense of balance (though you would learn much faster through expert commentary on your actions). Critical thinkers have a knowing *how* that is associated with all such experiential knowledge (Ellerton, 2015). This knowing how that critical thinkers have is not just about, for example, solving problems in their domain, it is also about making their thinking itself while problem solving an object of study. Critical thinkers know *how* to think, it is the quality of their reasoning that they are "critical" about and about which they care.

Thinking about thinking is a necessary characteristic of a critical thinker, but that does not mean that critical thinkers can only be effective when they are thinking about their thinking, or that they spend their days in this mode. It simply means that they are capable of doing so when required, and are capable of determining when this is required. To think about thinking is to be metacognitive, and it is in this context of critical thinking that we see the educational value of metacognition and of learning experiences that promote this.

We ask many things of aspiring critical thinkers, including that they seek and develop alternative perspectives and courses of action, generate useful inquiry questions, identify relevant and significant issues, express ideas with clarity and precision, give and demand reasons for accepting ideas or positions, evaluate claims and arguments according to their credibility and logical coherence and so on. But it is not enough to do these things without attention to how well they are being done or why doing them matters. For example, it is possible to seek alternative perspectives but to not go past the obvious ones; generate useful inquiry questions but not many of them; identify some relevant and significant issues but not be able to explain very well why they are relevant and significant; express some ideas with clarity but not others, produce reasons for a position but not very good ones, and so on. Critical thinking skills are not binary, they are done with a measure of success and applied with varying levels of mastery. Like all such things, they are improved with practice and with the right feedback. This leads us to two very important questions for educational contexts. The first is "what kind of activities allow students the opportunities to practice these thinking skills?" and the second is "how can we give feedback on the quality of student thinking so that it might be improved?"

In developing student's critical thinking skills, we must ensure that they have opportunities for doubt, for doubt is the reason for inquiry and inquiry provides a reason for valuing critical thinking. But doubt is not something to be allocated by the teacher, it must be cultivated.

> If [...] thinking in the classroom is considered desirable, the curriculum cannot present itself as clear and settled, for this paralyses thought. (Lipman 2003, p. 21)

The kinds of activities that best provide students with opportunities for developing their critical thinking skills are, therefore, those in which there is doubt about the outcome. In other words, the path of inquiry is not laid down for them to follow. This may not be the normal classroom experience for some students, since they spend much of their time trying to guess what response is expected of them and particularly so in classrooms with a strong focus on knowledge transfer, which can include STEM subjects. Activities that move students away from this mode of thinking and allow free inquiry, however, also demand attention to how students are thinking and how this thinking can be evaluated and justified as effective inquiry. Some examples of broad principles that may apply to the design and implementation of such activities in a STEM context are given in Table 2.1, below. These examples are necessarily few and brief, but they show how critical thinking can be enabled by the use of directed pedagogical strategies based on the need to engage with doubt within inquiry. These strategies also offer opportunities for feeding back to students on the quality of their thinking.

Opportunities for thinking need to be planned for in as much as detail and with as high a resolution as any as any lesson focused on presentation and transfer of discipline content. Such planning is necessary because thinking should not be an educational by-product of discipline instruction. We must also concern ourselves with what we must do in the classroom to develop the dispositions, habits and virtues of critical thinkers. It is important to appreciate that these are developed and formed through the business of inquiry, for an appreciation of what makes for good inquiry is "intimately intertwined" with a growth in virtue (Bailin & Battersby, 2016, p. 369). The kinds strategies that develop skills and dispositions, and for which the above are examples, are also those that, with pedagogical guidance and a metacognitive focus, help to develop inquiry virtues.

In considering the nature and contexts of critical thinking development, two significant and closely related aspects of teaching critical thinking can be further developed. The first is the need for collaborative inquiry, to give and receive feedback and to mediate the norms of inquiry, and the second the value of creativity to critical thinking.

2.5 Collaborative Inquiry

We have seen above the necessity for and characteristics of collaborative creativity, and some of the rewards of collaborative thinking are shown in Table 2.1. But thinking collaboratively is more than simply exchanging the outputs of our individual cognitive processes, it involves sharing the process of cognition itself. The kind of collaboration that is so productive in both creativity and critical thinking is of this sort. As discussed earlier in this chapter regarding creativity, this involves "a shift from ego-centric to group-centric dispositions of group members" and "high interactivity communication characterised by open-mindedness, empathy, flexibility, active and deep listening, respect and honesty [some key dispositions and virtues]".

Table 2.1 Pedagogical strategies and associated critical thinking opportunities

Enabling strategy	Potential critical thinking outcomes
Habitually seek the grounds of knowledge claims (how do/can we know that?). e.g. the universe is over 13 billion years old. How can we know that? How do we know that?	● Assumptions are tested and challenged. ● The limits of knowledge, potential and actual are recognised. ● Acceptance that knowledge, including scientific knowledge, is fallible. ● Development of the disposition to question knowledge claims.
Students problematise a situation (how can we best frame this problem?) rather than receiving a framework for investigation. e.g. is a drug epidemic a law and order problem or a medical problem?	● Alternative framings and perspectives are sought and found. ● Potential framings and consequent solutions are evaluated and justified.
Students determine the criteria for success e.g. who was Australia's greatest scientist?	● Assumptions are tested and challenged. ● Possibilities are considered and alternatives are generated. ● Possible positions and courses of action are evaluated and justified.
Students suggest what might be necessary knowledge and why. e.g. what would you like to know about a lake to explain the types of crystals that form around its edge? Why would you like to know that information? How do you think that knowledge could help explain the crystal structure?	● Students demonstrate their thinking regarding causal relationships and conceptual understandings. ● Students move beyond 'fishing' for information towards seeking causal relationships and more complex conceptual understandings. ● Schematically organised deep knowledge structures are developed.
Student investigative design. e.g. what materials do you think you would need to determine the factors that affect the period of a pendulum? Why do you think you will need those materials?	● Evidence is identified and data manipulation and processing is justified. ● How, when and why data is to be collected is explained in detail with relation to identifying/postulating causal relationships between variables.

(continued)

Table 2.1 (continued)

Enabling strategy	Potential critical thinking outcomes
Collaborative classroom dialogue in small or large groups based around tasks and activities in which outcomes are uncertain and inquiry methods need to be developed.	● Thinking is tested against the reasoning of others. ● An understanding of how the norms of inquiry are formed through rational dialogue is developed. ● Students engage in social cognition. ● Collaboratively developed norms of inquiry are internalised for later individual use. ● A broader and deeper range of creative options to assist in inquiry are generated (questions, framings, possible challenges, solutions, etc.). ● Cognitive biases are mitigated against using the checks of collaborative reasoning. ● A metacognitive language that focuses on the structure and quality of thinking is developed and used.

There are at least five reasons to value collaboration in critical thinking.

1. Collaborative thinking allows students to understand and develop the norms of effective thinking. Learning to think well is analogous to learning a language. It is not something you can do in isolation. Just as for language, reasoning is better thought of as a social competence (Sperber & Mercier, 2012) rather than an individual faculty since a key aspect of rationality is developing shared meaning through rational discourse. Developing the norms of good reasoning, especially through developing the dispositions and virtues of inquiry, is socially mediated.
2. Thinking collaboratively is a form of social cognition, in which the limits of cognition are not those of the individual mind but of the group. This can be understood in two ways, first that social cognition helps check the biases and assumptions held by individual minds, second that the outputs of one mind can act as inputs for other minds so forming a greater cognitive complex.
3. Creativity is a core component of critical thinking and this is best done collaboratively, as we have discussed above. Creativity is a process that demands critical analysis and evaluation and shares with critical thinking the need for (to revisit Guilford) fluency, flexibility and originality of thought, the ability and dispositions to reinterpretation and challenge old ideas and to move forward in the face of ambiguity. Without the ability to do this these things, thinking is limited to learned behaviours and patterns. Even though these patterns and behaviours might be quite sophisticated, they will not represent critical thinking.
4. Feedback is a necessary condition for improving experiential knowledge (knowing *how*). In collaborative inquiry, in which thinking is shared and communication is clear and direct, the interactions between participants provide opportunities

for immediate and frequent feedback on the appropriateness and quality of student thinking, either from their peers or from the teacher.

5. Collaborative inquiry forms and models the norms of effective thinking allowing development in critical thinking to be guided and shaped. The internalisation of what we learn socially becomes available as a resource for individual, private thinking (Vygotsky, 1978). In the broader sense, as Mead succinctly and originally puts it, "the child does not become social by learning. He [sic] must become social in order to learn" (1910, p. 693).

2.6 Creativity and Critical Thinking

We have discussed development in creativity in terms of collaborative and communicative growth, intrinsic motivation, generative capacity and complexity. The development of critical thinking—and of critical thinkers—is similar, though this growth is often expressed in the language of skills (and the handling of complexity), dispositions and habits (which map, in part, to intrinsic motivations) and virtues. Creativity and critical thinking are bound together by their mutual dependence and by their means of development. Creativity without criticality is rudderless and criticality without creativity is stagnant. The conditions under which both are developed include the presence of doubt (which might include doubt about both method and outcome), collaborative investigation, a shared commitment to determine truth or arrive at a solution, and an explicit instructional intent on the part of the teacher to develop both creativity and critical thinking together.

References

Altan, S., Lane, J. F., & Dottin, E. (2017). Using habits of mind, intelligent behaviors, and educational theories to create a conceptual framework for developing effective teaching dispositions. *Journal of Teacher Education, 70*(2), 169–183. https://doi.org/10.1177/0022487117736024.

Amabile, T. (2012). *The componential theory of creativity*. Boston: Harvard Business School.

Amabile, T., & Pratt, M. (2017). The dynamic componential model of creativity and innovation in organizations: Making progress, making meaning. *Research in Organizational Behavior, 37*, 157–183.

Annas, J. (1995). Virtue as a skill. *International Journal of Philosophical Studies, 3*(2), 227–243. https://doi.org/10.1080/09672559508570812.

Bailin, S., & Battersby, M. (2016). Fostering the virtues of inquiry. *An International Review of Philosophy, 35*(2), 367–374. https://doi.org/10.1007/s11245-015-9307-6.

Bowers, S. (2019). *Irish teenager wins Google science award for microplastics project*. Retrieved from https://www.irishtimes.com/news/science/irish-teenager-wins-google-science-award-for-microplastics-project-1.3971256

Dewey, J. (1938). *Logic: The theory of inquiry*. New York.

Dottin, E. (2009). Professional judgment and dispositions in teacher education. *Teaching and Teacher Education, 25*, 83–88. https://doi.org/10.1016/j.tate.2008.06.005

Ellerton, P. (2015). Metacognition and critical thinking: Some pedagogical imperatives. In M. Davies & R. Barnett (Eds.), *The Palgrave handbook of critical thinking in higher education* (pp. 409–426). https://doi.org/10.1057/9781137378057_25.
Facione, P. A. (1990). *Critical thinking: A statement of expert consensus for purposes of educational assessment and instruction. Research findings and recommendations*. Retrieved from http://www.eric.ed.gov/ERICWebPortal/detail?accno=ED315423
Gloor, P. (2017). *Swarm leadership and the collective mind*. Bingley, UK: Emerald.
Gotz, I. (1981). On defining creativity. *Journal of Aesthetics and Art Criticism, 39*, 297–301.
Guilford, J. (1959). Traits of creativity. In H. Anderson (Ed.), *Creativity and its cultivation* (pp. 142–161). New York: Harper.
IDEO. (2012). *Design thinking toolkit for educators* (2nd ed.). Retrieved from https://designthinkingforeducators.com
Kallick, B., & Costa, A. L. (2008). Learning and leading with habits of mind: 16 essential characteristics for success. *Association for Supervision & Curriculum Development*. http://ebookcentral.proquest.com/lib/uql/detail.action?docID=410671
Kelly, R. (2012). *Educating for creativity: A global conversation*. Edmonton: Brush Education.
Kelly, R. (2016). *Creative development: Transforming education through design thinking, innovation and invention*. Edmonton: Brush Education.
Kelly, R. (2020). *Collaborative creativity: Educating for creative development, innovation and entrepreneurship*. Edmonton: Brush Education.
Lipman, M. (2003). *Thinking in education* (2nd ed.). Cambridge, UK: Cambridge University.
Lubart, T. I. (2000). Models of creative process: Past, present and future. *Creativity Research Journal, 13*(3–4), 295–308.
Mead, G. H. (1910). The psychology of social consciousness implied in instruction. *Science, 31*(801), 688–693.
Mulnix, J. W. (2010). Thinking critically about critical thinking. *Educational Philosophy and Theory, 44*(5), 464–479. https://doi.org/10.1111/j.1469-5812.2010.00673.x.
OECD. (2018a). *Teaching, assessing and learning creative and critical thinking skills in education*. Retrieved from http://www.oecd.org/education/ceri/assessingprogressionincreativeandcriticalthinkingskillsineducation.htm
OECD. (2018b). *Fostering and assessing students' critical and creative thinking skills in higher education*. Retrieved from https://www.oecd.org/education/ceri/Fostering-and-assessing-students-creative-and-critical-thinking-skills-in-higher-education.pdf
Osborn, A. (1963). *Applied imagination*. New York: Charles Schribner.
Partnership for 21st Century Learning (P21). (2018). Retrieved from http://www.p21.org/members-states/partner-states
Paul, R., & Elder, L. (2008). *The miniature guide to critical thinking: Concepts and tools / by Richard Paul and Linda Elder* (5th ed.). Dillon Beach, CA: Foundation for Critical Thinking.
Piirto, J. (2004). *Understanding creativity*. Scottsdale: Great Potential.
Plucker, J., Beghetto, R., & Dow, G. (2004). Why isn't creativity more important to educational psychologists? Potentials, pitfalls, and future directions in creativity research. *Educational Psychologist, 39*, 83–96.
Sawyer, R. K. (2012). *Explaining creativity: The science of innovation*. New York: Oxford University.
Scriven, M., & Paul, R. (2011, November 23). *Defining critical thinking*. Retrieved from http://www.criticalthinking.org/pages/defining-critical-thinking/766
United Nations. (2019). *United Nations sustainable development goals*. Retrieved from https://www.un.org/sustainabledevelopment/
Siegel, H. (1989). Epistemology, critical thinking, and critical thinking pedagogy. *Argumentation, 3*(2), 127–140. https://doi.org/10.1007/bf00128144.
Siegel, H. (2017). *Education's epistemology: Rationality, diversity, and critical thinking*. https://doi.org/10.1093/oso/9780190682675.003.0007.

Sperber, D., & Mercier, H. (2012). Reason as a social competence. In H. Landemore & J. Elster (Eds.), *Collective wisdom—Principles and mechanisms* (pp. 368–392). New York: Cambridge University.
The Association of American Colleges and Universities. (2013). *It takes more than a major: Employer priorities for college learning and student success*. Retrieved from https://www.aacu.org/sites/default/files/files/LEAP/2013_EmployerSurvey.pdf
Topping, K. J., & Trickey, S. (2007). Impact of philosophical enquiry on school students' interactive behaviour. *Thinking Skills and Creativity, 2*(2), 73–84. https://doi.org/10.1016/j.tsc.2007.03.001.
Van Gelder, T., Bissett, M., & Cumming, G. (2004). Cultivating expertise in informal reasoning. *Canadian Journal of Experimental Psychology/Revue Canadienne de Psychologie Expérimentale, 58*(2), 142–152. https://doi.org/10.1037/h0085794.
Vygotsky, L. S. (1978). *Mind in society: The development of higher psychological processes*. Cambridge\London: Harvard University.
Waks, L. J. (2014). *Education 2.0: The learning web revolution and the transformation of the school*. Boulder: Paradigm.
Wallas, G. (1926). *The art of thought*. New York: Harcourt Brace.
Willingham, D. T. (2008). Critical thinking: Why is it so hard to teach? *Arts Education Policy Review, 109*, 21–32. https://doi.org/10.3200/AEPR.109.4.21-32.
Willingham, D.T. (2019). *How to teach critical thinking*. Retrieved from https://education.nsw.gov.au/media/exar/How-to-teach-critical-thinking-Willingham.pdf

Chapter 3
Fostering Students' Creativity and Critical Thinking in Science Education

Stéphan Vincent-Lancrin

Abstract What does it mean to redesign teaching and learning within existing science curricula (and learning objectives) so that students have more space and appropriate tasks to develop their creative and critical thinking skills? The chapter begins by describing the development of a portfolio of rubrics on creativity and critical thinking, including a conceptual rubric on science tested in primary and secondary education in 11 countries. Teachers in school networks adopted teaching and learning strategies aligned to the development of creativity and critical thinking, to these OECD rubrics. Examples of lesson plans and pedagogies that were developed are given, and some key challenges for teachers and learners are reflected on.

Keywords Creativity · Critical thinking · Science education · Innovation in education · Rubrics · Lesson plans

3.1 Introduction

What does it mean to redesign teaching and learning within existing science curricula (and learning objectives) so that students have more time and appropriate tasks to develop their creative and critical thinking skills?

The first difficulty to overcome is to operationalise the concepts of creativity and critical thinking so that each would be tangible and visible for science teachers. What do creativity and critical thinking mean in science education? What do these mean when students are not yet experts in their domain? To answer these questions, the OECD developed a portfolio of rubrics on creativity and critical thinking through

The analyses given and the opinions expressed in this chapter are those of the author and do not necessarily reflect the views of the OECD and of its members.

S. Vincent-Lancrin (✉)
Directorate for Education and Skills, Organisation for Economic Co-operation and Development, Paris, France
e-mail: stephan.vincent-lancrin@oecd.org

A. Berry et al. (eds.), *Education in the 21st Century*,
https://doi.org/10.1007/978-3-030-85300-6_3

a quick prototyping model, including a conceptual rubric on science that was tested in primary and secondary education in 11 countries between 2015 and 2019 (Vincent-Lancrin et al., 2019).

This chapter will show how teaching and learning strategies aligned to the development of creativity and critical thinking – and to those rubrics – could be used in science education. The first section will present the rubrics and how they related to theories of creativity and critical thinking in general. The second section will present two examples of lesson plans and pedagogies that were developed during the OECD project to foster and assess creativity and critical thinking in education. Beyond the above-mentioned rubrics, these lesson plans illustrate how the development of creativity and critical thinking skills can look in practice in a science unit. The chapter will also reflect on some key challenges for teachers and learners to make the development of creative and critical thinking skills possible. It will conclude by highlighting the importance of integrating similar approaches in other school subjects so that students experience enough opportunities to develop those skills.

3.2 How to Support Creativity and Critical Thinking in Science Education: Concepts and Rubrics

Most contemporary education systems include creativity and critical thinking as part of their list of key skills students should acquire in their schooling. Most curricula in OECD countries do include in one form or another critical thinking and creativity as expected learning outcomes. Their importance in education and higher education has become consensual worldwide (Fullan, Quinn & McEachen, 2018; Newton & Newton, 2014; Lucas & Spencer, 2017). The role of education in the development of critical thinking is also increasingly acknowledged within many countries, where a majority of the population believe that schools should help students to become "independent thinkers" rather than passive receivers of transmitted knowledge – or at least recognise the importance of such an objective (Fig. 3.1). Developing critical thinking and creativity leads to more independent thinking, which can thus be considered as a good proxy for those skills .

However, even though the importance of creativity and critical thinking is usually well accepted, it remains unclear to teachers what these terms actually mean and entail in education. In order to create a shared professional language on creativity and critical thinking in education, the OECD worked over five years with a network of schools and teachers in 11 countries (Vincent-Lancrin et al., 2019). (The countries are: Brazil, France, Hungary, India, the Netherlands, Slovak Republic, Russian Federation, Spain, Thailand, United States, United Kingdom [Wales].)

In addition to the lack of clarity on the definitions of those skills, another difficulty lies in the levels of teacher-friendliness of the language used. To this effect, a portfolio of rubrics was developed to help teachers be more informed, intentional

Answer to the question: "It is more important that schools in our country teach…"

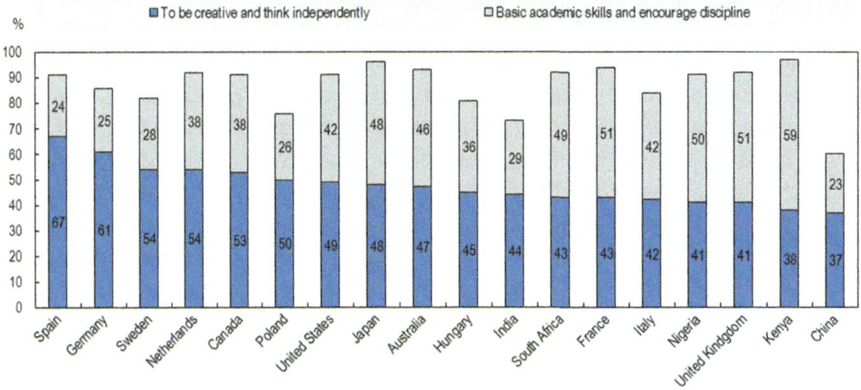

Fig. 3.1 Many societies support the fostering of creativity and critical thinking in education. (Source: Pew Research Centre, Spring 2016 Global Attitudes Survey)

and consistent in their efforts to develop their students' creativity and critical thinking. A conceptual rubric for science education is part of this portfolio.

3.2.1 Creativity and Critical Thinking

Creativity and critical thinking are two distinct but related higher-order cognitive skills. As such, both require significant mental effort and energy; both are cognitively challenging. Creativity aims to create novel, appropriate ideas and products. Critical thinking aims to carefully evaluate and judge statements, ideas and theories relative to alternative explanations or solutions so as to reach a competent, independent position – possibly for action.

The research on creativity and research on critical thinking actually do not overlap much, even though critical thinking often plays an important role in creativity, and vice versa (see Ellerton & Kelly, Chap. 2). School curricula and educational rubrics are however prone to group the two together and to talk about "creative and critical thinking". In the same spirit, Lucas and Spencer (2017) include critical thinking (as well as problem solving) under the concept of "creative thinking".

Sternberg and Lubart (1999) proposed a simple definition of creativity: "creativity is the ability to produce work that is both novel (i.e., original, unexpected) and appropriate (i.e., useful, adaptive concerning tasks constraints" (p. 3). The use of "appropriate" in this definition reminds us that creativity happens within a system or context with its established standards; it is not just about doing something new. As Dennett (2013) puts it: "Being creative is not just a matter of casting about for something novel – anybody can do that, since novelty can be found in any random juxtaposition of stuff – but of making the novelty *jump* out of some *system*, a system that has become established, for good reason" (p. 45).

Emphasising both process and output, Lubart (2000) defines creativity as "a sequence of thoughts and actions that leads to novel, adaptive production" (p. 295). What is this sequence? Creativity research has explored the cognitive processes involved in creativity. Guilford (1950) emphasised two processes leading to creativity: *divergent thinking* (generating many ideas) and *convergent thinking* (choosing and developing a good one). Torrance (1970), distinguished four aspects of the creativity process: *fluency* (having many relevant ideas), *flexibility* (having different types of relevant ideas), *originality* (having statistically novel ideas) and *elaboration* (being able to elaborate one's ideas). Most standardised tests of creativity or creative potential (e.g., Torrance, Wallach-Kogan, Guilford, Getzel-Jackson, Mednick, Runco) decompose the creative process along similar lines and focus on some of its aspects.

Critical thinking may be a step in the creative process, or may not: convergent thinking does not necessarily have to be "critical" (Runco, 2009). Critical thinking mainly aims at assessing the strength and appropriateness of a statement, theory or idea through a questioning and perspective-taking process – which may in turn result (or not) in a possibly novel statement or theory. Critical thinking need not lead to an original position to a problem: the most conventional one may be the most appropriate. However, it typically involves the examination and evaluation of different possible positions.

In education (including higher education), the theory of critical thinking has been developed by philosophers such as Ennis (1996, 2018), Facione (1990) and McPeck (1981) (see Davies & Barnett, 2015, and Hitchcock, 2018, for overviews of the literature). Hitchcock (2018) summarises most conceptions by defining critical thinking as "careful goal-directed thinking" – another version of Ennis' definition: "reasonable reflective thinking focused on deciding what to believe or do" (Ennis, 2018, p. 165). In many cases, definitions of critical thinking emphasise logical or rational thinking, that is, the ability to reason, assess arguments and evidence, and argue in a sound way to reach a relevant and appropriate solution to a problem. However, critical thinking also includes a dimension of "critique" and "perspective-taking". In addition to rational or logical thinking, critical thinking thus includes two other dimensions: the recognition of multiple perspectives (and/or the possibility of challenging a given one) and the recognition of the assumptions and limitations of any perspective, even when that perspective appears superior to all other available ones.

Many of the cognitive processes involved in creativity and critical thinking share commonalities. Both require prior knowledge in the domain of application. The sub-skills that need to be deployed for each skill involve imagining, inquiring, doing and reflecting. Creativity puts more emphasis on imagining (brainstorming, generating ideas and alternatives), while critical thinking places more emphasis on "inquiring", including its more analytical and systematic dimensions (understanding and decomposing the problem, etc.). Critical thinking is primarily inquisitive, a detective way of thinking; creative thinking is more imaginative, an artist way of thinking. However, critical thinking does involve imagining alternative theories, counterfactuals, reasons, and results in an action (making a judgment); creativity does require making judgments and decisions about the alternative ideas generated in the imaginative process, and, more fundamentally, the examination of the assumptions of existing solutions and conventions. In this sense, creativity and critical thinking can be thought of as two ends of a continuum.

Both creativity and critical thinking require a certain level of openness and curiosity. Both may lead to challenges to authority, values or accepted norms; this is what may make them both valuable, and sometimes challenging. Critical thinking requires integrity; creativity requires discipline and judgment. When education is conceived as the mere transmission of socially accepted knowledge, there is little room for either. In fact, like most other skills, creativity and critical thinking only have to be exercised at some points; even if a world in which people would be creative all the time or critical all the time was concretely possible this world would be most dysfunctional. Students also need to learn when and about what they can or should think creatively or critically. In an educational context, both creative and critical thinking necessarily pursue the deeper understanding of knowledge and solutions, and thus deeper learning. Developing creativity and critical thinking is actually a way to improve learning and achievement – whether such thinking leads to the proposing of new knowledge and solutions or not.

Even though one can describe them at the conceptual level in a domain-general way, both creativity and critical thinking in practice are mainly domain-specific: each requires knowledge about a field or context to be practiced, and usually being a strong creative or critical thinker in a particular domain does not imply any transfer of those skills to another domain. The research literature overwhelmingly emphasises the "domain-specificity" of both, even though at the conceptual level each can be described in a domain-general way.

3.2.2 Rubrics to Support Creativity and Critical Thinking in Science Teaching and Learning

There is overall a common understanding among researchers on the key dimensions of creativity and of critical thinking. However, transferring the concepts to a consistent educational application requires further translation. This is where rubrics intervene. Rubrics are a way to simplify, translate and construct social representations of what creativity and critical thinking look like in the teaching and learning process, and so create a shared understanding of what each means in the classroom, and lead to common expectations among teachers, and among teachers and students. The function of rubrics is to simplify and elaborate the complex concepts of creativity and critical thinking so that they become relevant to teachers and learners in their actual educational activities. The rubrics also allow teachers to monitor and formatively assess whether their students develop those skills. Rubrics are a metacognitive tool that helps make learning visible and tangible, and teaching intentional.

Different types of rubrics serve different purposes. "Conceptual rubrics" are those that clarify "what counts" or "what teachers and students should particularly keep in mind", while "assessment rubrics" articulate levels of progression or proficiency involved in the acquisition of creative and critical thinking skills. Both types were developed in the OECD project from which this chapter draws, here we will focus on only the conceptual rubrics.

The development of rubrics requires balancing between simplicity and complexity. To be useful for teachers and classrooms, rubrics have to be teacher-friendly (and possibly student-friendly), and have a language that is easily understandable by teachers at different school levels. On the one hand, the descriptors of the different key ideas have to relate sufficiently to the concepts as understood by experts in creativity and critical thinking. On the other hand, the descriptors have to be simple enough to be easily understood by teachers and students, and have to relate to skills and activities that are meaningful in school settings. Ideally, one would easily memorise some of the language used in the rubric so that this becomes internalised. Using a language inspired by the "five habits of mind" rubric developed by Lucas, Claxton and Spencer (2013), and a review of other existing rubrics, the OECD rubrics that are the focus in this chapter tried to capture different dimensions of both creativity and critical thinking through four high level and easily memorable descriptors (dimensions): *imagining*, *enquiring*, *doing*, *reflecting*. Each of those active words is then associated with some more specific descriptor(s) for creativity and for critical thinking.

Two domain-general conceptual rubrics were developed: a "comprehensive" rubric and "class-friendly" rubric. Domain-specific adaptations of those rubrics were also developed, including for science. Table 3.1 shows the "comprehensive" domain-general rubric, while Table 3.2 presents the "class-friendly" rubric for creativity and critical thinking in science education.

In the case of creativity, the four dimensions in the left hand column of Table 3.1 can be elaborated as follows:

- *Inquiring*. This dimension of the creative cognitive process is close to scientific inquiry. Torrance (1966) highlights the importance of identifying problems, gaps in knowledge, missing knowledge and elements in the creative process. Because creativity cannot happen without knowledge about the field or problem investigated, looking for information, finding the problem and understanding its different possible dimensions are important aspects of the creative process. These can take different forms, depending on the problem, from feeling and empathising with people to a more objective approach of observing, describing and analysing from different possible perspectives what the issues and problems at stake are. Both curiosity and unconventional connections between different knowledge and problems matter in the creative inquiry process.
- *Imagining*. Imagination refers to the ability to see and play with ideas and things in one's mind. This ability allows people to get free from conventional reality

Table 3.1 OECD rubric on creativity and critical thinking (domain-general, comprehensive)

	CREATIVITY (Coming up with new ideas and solutions)	CRITICAL THINKING (Questioning and evaluating new ideas and solutions)
INQUIRING	• Feel, empathise, observe, describe relevant experience, knowledge and information	• Understand context/frame and boundaries of the problem
	• Make connections to other concepts and ideas, integrate other disciplinary perspectives	• Identify and question assumptions, check accuracy of facts and interpretations, analyse gaps in knowledge
IMAGINING	• Explore, seek and generate ideas	• Identify alternative theories and opinions and compare or imagine different perspectives on the problem
	• Stretch and play with unusual, risky, or radical ideas	• Identify strengths and weaknesses of evidence, arguments, claims and beliefs
DOING	• Produce, perform, envision, prototype a product, a solution or a performance in a personally novel way	• Justify a solution or reasoning on logical, ethical or aesthetic criteria/reasoning
REFLECTING	• Reflect and assess the novelty of chosen solution and of its possible consequences	• Evaluate and acknowledge the uncertainty or limits of the endorsed solution or position
	• Reflect and assess the relevance of chosen solution and to its possible consequences	• Reflect on the possible bias of one's own perspective compared to other perspectives

Note: This rubric is intended for teachers/faculty use to identify the student skills related to creativity and to critical thinking that they have to foster in their teaching and learning, not for assessment

and to pursue novel ideas and invent new stories, anticipate the future, pursue different scenarios, envision counterfactuals, simulate consequences of different ideas and solutions, etc. In the context of creativity, imagination is about a free and playful generation of ideas, theories and assumptions, with a certain level of intentionality. This can take the form of an independent generation of multiple ideas or association of ideas, either by seeing actual or sometimes metaphorical connections (Mednick, 1962; Runco, 2009). Being able to push ideas to their limits, or to explore unconventional (or even seemingly absurd) ideas without much actual risk, is one of the cognitive processes that creativity may involve.

- *Doing*. Creativity implies the creation of something novel and appropriate, based on one's inquiry and imagination. This is typically the convergent or integrative part of the creative process. This output production can take different forms based on the domain: it can be a product, a performance, an idea, a physical or mental model, etc. It implies the selection of some of the ideas that have been imagined and inquired, and thus some level of reflection and audacious decision-making to meet the two main aspects of creativity. While products can be associated with the final stage of the creative process, the creative process can also

Table 3.2 Class friendly rubric (Science)

	CREATIVITY (Coming up with new ideas and solutions)	CRITICAL THINKING (Questioning and evaluating new ideas and solutions)
INQUIRING	• Making connections to other scientific concepts	• Identify and question assumptions and generally accepted ideas of a scientific explanation or approach to a problem
IMAGINING	• Generate and play with unusual and radical ideas when approaching or solving a scientific problem	• Consider several perspectives on a scientific problem
DOING	• Pose and propose how to solve a scientific problem in a personally novel way	• Explain both strengths and limitations of a scientific solution based on logical and possibly other criteria (practical, ethical, etc.)
REFLECTING	• Reflect on steps taken to solve a scientific problem	• Reflect on the chosen scientific approach or solution relative to possible alternatives

Note: This rubric identifies the main relevant subskills related to creativity and critical thinking that students should develop as part of their science education. It is not meant to score students or provide them with a continuum of skill progression

include some tinkering processes of trial and error, or the development of prototypes and models, and can intervene at different stages of the process.

- *Reflecting*. Finally, intentionality and reflection are key aspects of creativity. Intentionality distinguishes creativity from random novelty, and sometimes from small children's spontaneity. The level of intentionality and reflection can vary with age, but also with one's level of creative proficiency. As noted above, reflection also occurs at different stages of the creative process as one decides which ideas to select and how to move forward.

While these different aspects of creativity do not necessarily come in a definite order, or are solicited at different points in the creative process, the four can easily be related to the design thinking method, which codifies the innovation or creativity process and aims to turn it into an art (Kelley, 2001; see Kelly and Ellerton, Chap. 2). For educational purposes, the d.school at Stanford University summarised the innovation process in five steps that can be looped: empathise, define, ideate, prototype, test. Many of those processes are included in the proposed rubrics.

In the case of critical thinking, in order to have a parallelism with creativity, the underlying cognitive processes or sub-skills can be described under the rubrics' headings:

- *Inquiring*. Determining and understanding the problem at hand, including its boundaries, is a first important dimension of critical thinking's inquisitive process. Sometimes this includes wondering about why the problem is posed in a certain way, or examining whether the associated solutions or statements may be based on inaccurate facts or reasoning and identifying the knowledge gaps. This inquiry process partly concerns rational thinking (checking facts, observing, analysing the reasoning), but also includes a more "critical" dimension when it comes to identifying the possible limitations of the solution and challenging some of the underlying assumptions and interpretations, even when facts are accurate. In many cases, inquiring involves acquiring knowledge, verifying knowledge, and examining the components of the problem in detail as well as the problem as a whole.

- *Imagining*. In critical thinking, imagination plays an important role as the mental elaboration of an idea – but any thinking involves some level of imagination. At a higher level, imagining is also about identifying and reviewing alternative, competing world views, theories and assumptions, so as to consider the problem from multiple perspectives. This allows for a better identification of the strengths and weaknesses of proposed evidence, arguments and assumptions, even though this evaluation also belongs to the inquisitive process. Imagination also plays a role in thought experiments, which can be a strong component of any good thinking and also a way to make a point when experimentation is not possible (Dennett, 2013).
- *Doing*. The product of critical thinking is one's position or solution to a problem (or judgment about others' positions or solutions). This mainly implies careful inference, a balancing act between different ways of looking at the problem, and thus recognition of its (possible) complexities. As in any productive thinking, critical thinking implies the ability to argue and justify one's position rationally, according to some existing perspectives and socially recognised ways of reasoning, or possibly some new ones.
- *Reflecting*. Finally, even though one may consider one's position or way of thinking superior to some alternatives, perhaps just because it embraces a wider view or is better supported by existing evidence, critical thinking implies some self-reflective process about the perspective one endorses, its possible limitations and uncertainties, and thus a certain level of humility and openness to other competing ideas. While one does not have to embrace ancient scepticism and suspend one's judgment in all cases, this may sometimes be the most appropriate position.

The OECD rubrics for creativity and critical thinking were meant to be used by teachers working in real-life settings in different ways: (1) designing and revising lesson plans so that they would give students the opportunity to develop their creativity and critical thinking skills; (2) assessing student work and progression in the acquisition of these skills; (3) generating new aligned rubrics adapted to their local context or self-assessment tools. Field work showed that seven in ten teachers participating in the international network did on average use the OECD rubrics for those purposes. The rubrics have thus proven to be useful and well adopted by teachers in most of the countries in which the project was implemented.

3.2.3 Creativity and Critical Thinking in Science

While science education can be one of the many vehicles to develop students' creativity and critical thinking in a school context, it is noteworthy that critical thinking and creativity are also at the core of scientific practice. When practiced by expert scientists, science is about creativity and critical thinking.

Scientists usually need to have creative or original ideas to receive grants and get published in scientific journals. Scientific awards (such as the Nobel Prizes) typically celebrate advances that bring some ideas or techniques that are "new to the world" (and in this sense, creative in the full meaning of the word). One aspect of scientific practice that is usually somewhat downplayed is "imagination". It is nevertheless a key aspect of science as Nobel Prize winner and famous physicist Feynman (1963) noted:

> *Experiment is the sole judge of scientific "truth."* But what is the source of knowledge? Where do the laws that are to be tested come from? Experiment, itself, helps to produce these laws, in the sense that it gives us hints. But also needed is imagination to create from these hints the great generalizations—to guess at the wonderful, simple, but very strange patterns beneath them all, and then to experiment to check again whether we have made the right guess. This imagining process is so difficult that there is a division of labor in physics: there are theoretical physicists who imagine, deduce, and guess at new laws, but do not experiment; and then there are experimental physicists who experiment, imagine, deduce, and guess. (p. 1)

As for critical thinking, it is in fact at the heart of scientific progress – and one of the prerequisites of science. Science's very core value is doubt, the possibility to question what authorities (including teachers and scientists) say. There is no science without a certain level of scepticism, as Feynman (1955) forcefully noted:

> The scientist has a lot of experience with ignorance and doubt and uncertainty, and this experience is of very great importance [...] We have found it of paramount importance that in order to progress we must recognize our ignorance and leave room for doubt. Scientific knowledge is a body of statements of varying degrees of certainty - some most unsure, some nearly sure, but none absolutely certain. [...] Our freedom to doubt was born out of a struggle against authority in the early days of science. It was a very deep and strong struggle: permit us to question - to doubt - to not be sure. I think that it is important that we do not forget this struggle and thus perhaps lose what we have gained. Herein lies a responsibility to society. [...] It is our responsibility as scientists... to teach how doubt is not to be feared but welcomed and discussed; and to demand this freedom as our duty to all coming generations. (pp. 245–247)

Teaching and learning creativity and critical thinking in science education in schools is thus one way to "think like a scientist" and understand the values of science, even if, as for the technical skills in science (that is, the mastery of content and procedural knowledge), students are not necessarily expected to be as proficient as expert scientists – not to mention the most celebrated ones.

3.3 Creativity and Critical Thinking in Action in Science Education

Depending on the subject of the lesson and the learning outcomes they want to achieve, using a conceptual rubric while designing a lesson helps teachers to build in some assignments or tasks giving students the opportunity to develop at least some of the sub-skills of creativity or critical thinking. Some lessons may aim to develop just a few sub-skills, while others could cover the full range, with an emphasis on either creativity or critical thinking (or both). Existing lessons could be modified according to the same process, just adding one opportunity to develop a sub-skill here and another there through small changes to the lesson or its pedagogical delivery.

The conceptual rubrics also represent a key element of a quality assurance method: after decomposing their lessons or entire course into steps, teachers can identify when students were given the possibility or were requested to practice some of the skills identified in the rubric. Examples of lesson plans developed during the lessons/ course can thus include a mapping of the different steps of the lesson against the sub-skills of the conceptual rubrics.

Teams working on redesigning their science education courses implemented the OECD project that is the focus of this chapter in different ways. Two "signature pedagogies" were used by some of the teams (project-based and research-based learning), while most others just designed short projects or activities or improved more traditional lesson plans.

One example of lesson plans in science education grounded in project-based learning, and included in the OECD examples of courses, is now presented to give tangible ideas of how critical thinking and creativity can be developed in science education while also teaching technical skills of science (declarative and procedural knowledge).

3.3.1 What Controls My Health?

Developed by Adler et al. (2017), "What controls my health?" is a 20-lesson course engaging students in investigations to understand the importance of both genetic and environmental factors in their risk for disease. Students start the unit by experiencing the phenomenon of Type 2 diabetes through the eyes of a peer recently diagnosed with the disease. They develop an initial model to answer the driving question of the whole project: "What caused Monique's diabetes?" The driving question is particularly relevant to the students for whom it was designed and who live in Detroit, a city which is predominantly African-American and where most students are likely to have relatives suffering from diabetes.

Throughout the unit, students learn that diabetes, like many common diseases, is caused by a combination of both genetic and environmental factors. They also investigate how lifestyle options for healthy foods and exercise help prevent or reduce Type 2 diabetes. One lesson includes several opportunities for students to construct, test, revise and share their models to explain the investigated phenomena, while performing experiments and using computer simulations. For their final assignment, students conduct an action research project, based on their scientific and technological knowledge and understanding, which aims to improve the health of their school or neighbourhood to help prevent or reduce diabetes.

A summary description of the course is now presented (a more elaborated outline is publicly available at Adler et al., 2017):

1. *Periods 1–2: Why does Monique have diabetes?* Students learn about Type 1 and Type 2 diabetes (video). They develop an initial model that explains a health phenomenon of their choice.
2. Periods 3–5*: How can we describe Monique's diabetes?* Students learn more (through reading), and share information about the cause, symptoms and treatment of both Type 1 and Type 2 diabetes. They perform a glucose tolerance test by analysing simulated blood plasma samples to determine if the person has Type 1 or Type 2 diabetes. They learn about the heart, as an example of an organ which may be affected by diabetes. They revisit the *Driving Question Board*[1] and reflect upon their learning. They revise their models and add the biological aspect of diabetes to their model.
3. Periods 6–9: *How does Monique's family affect her diabetes?* Students examine pictures of a family to identify some genetic factors of characteristics that might be inherited. They collect data on tongue rolling and arm span, and use these data to explore the population variation of the inheritance patterns of single and multi-factorial genes. They use beads to simulate the inheritance of risk factors for diabetes. They identify the risk of diabetes in offspring based on the number and type of risk factors inherited during the simulation. They revisit the *Driving Question Board* and reflect upon their learning. They revise their models and add the effect of genetic factors on Monique's diabetes.
4. Periods 10–12: *How does where Monique lives and what she does affect her diabetes?* Students study the influence of environment on living organisms through plant growth.
5. Periods 13–16: *How do Monique's characteristics and environment affect her diabetes?* Through simulation, students consider how genetics and environment affect the health of sand rats.
6. Periods 17–18: *What can Monique do to make her environment healthier?* Students study the role of nutrition.
7. Periods 19–20: *Community action projects: How can we work together to make our environment healthier?* Students develop and choose their inquiry question, design and develop their research tools, then plan and carry out their investigations. They analyse the data and draw conclusions, share their findings with their

[1] Many project-based science units/ courses initially develop "Driving Questions" to contextualise the unit and give learners opportunities to connect the unit to their own experiences and prior ideas.

peers and broader community, suggest solutions and potential actions based on their findings.

This sequence is a good example of how teachers could allow their students to learn about science technical skills while giving them opportunities to also develop their creativity and critical thinking (as well as some social and behavioural skills). In terms of technical skills, that is, the mastery of content and procedural scientific knowledge, students clearly learn about: diabetes; the heart as an organ; the growth of plants; genetics, the influence of environmental factors; nutrition; the multiple drivers of health; making tests and experiments, including through computer simulation, and interpreting them.

The main focus of the "What controls my health?" course is actually critical thinking as students: identify and question their assumptions or accepted ideas about diabetes and its causes (steps 1 and 7 above); consider several perspectives on the problem at hand (steps 3 to 6); explain both the strengths and limitations of their scientific solution (steps 6 and 7); and consistently reflect on the chosen scientific approaches that they consider relative to possible alternatives (steps 2, 3, 4 and 7).

The lessons also allow students to develop some creativity skills as they: are induced to make connections to other scientific concepts or ideas throughout the project and to use remote examples to better understand (heart, plants) (steps 2 and 5); generate and play with unusual ideas as they revisit the questions of the Driving Question Board and have to generate their own solution (steps 1, 4, 7); propose how to solve a scientific problem in a personally novel way (steps 1 and 7); and reflect on those steps at the end of the process (step 7).

3.3.2 Evaporative Cooling

Another 10-lesson science (chemistry) unit was developed as part of a US-Finland project on problem-based learning in science, showing how to craft optimal learning moments in science learning environments (Schneider et al., 2020). The unit was also contributed to the OECD bank of pedagogical resources and is called "Evaporative cooling" (Paddock et al., 2019). It engages students in investigating the following driving question: "when I am sitting by the pool, why do I feel colder when I am wet than when I am dry?"

Students learn about intermolecular forces and energy transfer between molecules during phase changes of matter. Students start by experiencing the phenomenon of evaporation and cooling of different liquids. Later activities include experiments to measure temperature and mass changes when liquids evaporate, and, using several computer-based simulations, to explore energy transfer, forces, and interactions between molecules in different phases. Throughout, and under the guidance and with the scaffolding of their teacher, students continuously build, use, evaluate, and revise their own computational and hand-drawn models to answer the driving question.

A summary description of the course is now presented (a more elaborated outline is publicly available at Paddock et al., 2019):

1. Lesson 1: *Why does having wet skin makes you feel cooler?* Students are introduced to the Driving Question, and construct and draw their initial model of the phenomenon in pairs so that they access their prior knowledge about this phenomenon. The purpose for the teacher is for formative assessment to inform planning for lesson 2. Student pairs explain their model to another pair in turns so that they can share ideas and begin working toward some consensus in understanding.
2. Lesson 2: *Does evaporation depend on coverage?* Students go to two stations that are designed to test the rate of evaporation of acetone, water, and ethanol and the temperature change during the process. Students observe temperature change across time while the liquid is covered and while uncovered. Students observe mass change across time (covered and uncovered), collect data and graph these. Following the activity, students answer questions to guide their noticing of patterns.
3. Lessons 3–4: *Why does having wet skin makes you feel cooler?* (a specific return to the driving question) Students review their drawn models from lesson 1 and learn how to use a new modelling tool, SageModeler (freely available at https://learn.concord.org/building-models). They create an initial model of their experimental results using the modelling tool. Students use a computer simulation to compare properties of the states of matter. Properties include: spacing of molecules (which connects with potential energy) and kinetic energy.
4. Lesson 5: *How does thermal energy work?* Students learn that thermal energy can be transferred during phase changes. They develop a model to show how a system gains or loses thermal energy.
5. Lesson 6: *Why does having wet skin makes you feel cooler?* (a further deliberate return to the driving question) Students reflect on their drawn models. They review the learning from Lesson 5 and incorporate ideas from the lesson into their SageModeler model, and share their models to receive feedback.
6. Lessons 7–8: *Do matter viscosity and intermolecular forces play a role?* Students compare viscosity of water to acetone by observing how each chemical spreads on two different surfaces (a coin and wax paper). They investigate how the strength of the intermolecular forces in a substance effect the state of matter of a substance at a certain temperature using a computer simulation.
7. Lessons 9–10: *Why does having wet skin makes you feel cooler?* (a final return to the driving question) Students revisit their model, and, in pairs again, create a final draft of the model and evaluate this using a provided rubric. Students assess their peers' models with the rubric, and make a final revision of their own model. They present their final models and explanations to the whole class and may take a unit test. Students then perform a final assessment of the unit to reflect on what has been learned.

Here, again, there are clear technical skills to be acquired, both in terms of scientific content and procedural knowledge. In terms of content knowledge, students learn about the structure of matter, the behaviour of particles, intermolecular forces, the position and arrangement of atoms, and about relationships between molecular motion, the position of molecules, and kinetic and potential energy. The model they build and revise seeks to identify a pattern between the structure of particles and the behaviour (evaporation and temperature) of particles. In terms of procedural knowledge, students learn to model a phenomenon, to experiment, to analyse data, and to revise their model.

Although the sequence of lessons is essentially science (chemistry), it also gives students more opportunities to develop their creativity - and this more than their critical thinking. Or, to put it another way, the sequence does not dwell on identifying and questioning assumptions and generally accepted ideas of a scientific explanation or approach to a problem. The lesson does not offer students as many different perspectives as the previous example sequence ("What controls my health?"), as that was organised to show how different branches of science and technology explain different interacting parts of the puzzle. Nevertheless, "evaporative cooling" does seek to help students understand that *several* aspects of matter have to be factored in to understand and explain the phenomenon.

Students exercise their scientific creativity by repeatedly imagining how the phenomenon under study can be explained, generating and playing with different ideas that are unusual and new to them, making connections with their life but also with other knowledge they have. They are given opportunities to put forward an explanation of this scientific problem in a personally novel way at different stages of the unit (lessons 3, 5 and 8). They have to imagine from the outset how the phenomenon could be explained, making connections with their own and new experiences of evaporation, and as they look for and are given new information and knowledge, they continue to play with new ideas and imagine new solutions that they revise throughout the unit. Whether they have room to play with unusual or radical ideas depends on the teacher, who could very well encourage them to go in that direction to possibly prove them wrong or help them understand what a scientific (falsifiable) statement looks like.

In the Evaporative cooling lessons students practice some of their critical thinking skills by considering several perspectives on the driving question, gradually enriching and adding new concepts within the theoretical frame in which they operate. They have considerable room throughout the lessons to reflect individually and collectively on the strengths and limitations of the successive models they elaborate, identifying gaps and looking for alternative models.

The lessons shows that, even when a unit is framed around one main "theory" or "knowledge" to learn, there is room for students to have some level of agency, to imagine possible solutions, inquire about them, craft experiments and models to test their ideas, and reflect on them.

3.3.3 Design Criteria for Good Lessons

While the conceptual rubrics presented above can support teachers to review their curriculum units and plan lessons that give students opportunities to develop the sub-skills identified by the rubrics, they do not provide guidance on all key dimensions of the pedagogical work. In fact, while creativity and critical thinking can be nurtured in any domain, within and outside of science (or other apsects of STEM), these do require giving students certain types of tasks and problems. A set of "design criteria" was thus developed by Vincent-Lancrin et al. (2019) to support teachers further, building on learning science principles, including motivation, cognitive activation, self-regulation and opportunities for formative assessment (see Table 3.3). These design criteria for good lesson plans represent another set of

Table 3.3 Design criteria for activities that foster creativity or critical thinking skills

A pedagogical activity aligned with the OECD rubric on creativity and critical thinking should:	Comments
1. Create students' need/interest to learn	• Usually implies starting with a big question or an unusual activity. • May imply coming back to these questions several times during the activity.
2. Be challenging	• Often, the lack of student engagement comes from learning goals or activities that lack challenge. The tasks should be challenging enough, though not too difficult given the students' level.
3. Develop clear technical knowledge in one domain or more	• The activity should include the acquisition and practice of both content and procedural knowledge (technical knowledge).
4. Include the development of a product	• A;product (a paper, a presentation, a performance, a model, etc.) makes the learning visible and tangible. • Teachers and students should also be attentive to and possibly document the learning process.
5. Have students co-design part of the product/solution or problem	• Products should thus in principle not look all alike.
6. Deal with problems that can be looked at from different perspectives	• Problems should have several possible solutions. • Several techniques may be used to solve them.
7. Leave room for the unexpected	• Teachers and students do not have to know all the answers. • The most commonly adopted techniques/solutions may have to be taught and learnt, but there should be room for exploring or discussing unexpected answers
8. Include space and time for students to reflect and give/receive feedback	

Source: Vincent-Lancrin et al. (2019)

quality checks and new perspectives on how to approach pedagogical redesign to foster students' creativity and critical thinking.

The "design criteria" highlight that tasks to develop and then demonstrate creativity and critical thinking skills in education share some general features: they seek to engage students, they may have a deliberately open nature, and they encourage students to explore multiple solutions to problems within parameters and constraints that clarify goals yet remain relatively flexible to allow students to address them with a certain level of agency.

The successful teaching of creativity and critical thinking also hinges critically on teachers' attitude and in their ability to create learning environments where students feel safe to take risks in their thinking and expressions. This in turn presupposes a positive attitude towards mistakes and learner empowerment. A positive attitude among teachers towards student "mistakes" or "failure" can take the form of using these to trigger reflection about opportunities for learning, thus helping students to see misunderstandings and other matters too often labelled 'failures' as a chance for improvement (see Mansfield & Gunstone, Chap. 9). Choosing questions and tasks that teachers themselves cannot resolve can make it clear to students that the thinking process behind a problem can be as important as its answer. This is typically the role of the Driving Question Board in project-based learning (Schneider et al., 2020), something that demands a positive teacher attitude towards students' questions, and also students' explanations.

3.4 Concluding Remarks

To foster their students' creative and critical thinking skills in science education, teachers have to be intentional – and thus clear about what creativity and critical thinking mean in an educational setting, what subskills they should have their students practice, and what they should observe and monitor in the classroom. This clarity could typically be provided by the use of rubrics on creativity and critical thinking, both to create a more accessible and better understanding of what creativity and critical thinking entail and to ensure that students have opportunities to practice these higher order skills during their class work.

Exemplars of lesson plans or curriculum units should typically supplement those rubrics and illustrate how to equip students with creativity and critical thinking skills while teaching and learning traditional science education subjects. It is noteworthy that resources such as rubrics and examples of lesson plans are just a second-best but a much cheaper and much more widely available option than direct professional development of teachers can ever be.

While science and STEM education can provide tasks that would allow students to develop both their critical thinking and their creativity, the consistent acquisition of creative and critical thinking skills must be reinforced by other disciplines as well. There are two main reasons for this. The first is time. It takes practice to develop any skill, and it is possible that science lessons in school cannot provide enough occasions (hours) for students to practice the creativity and critical thinking skills that have been highlighted in this chapter. Significant reinforcement and development of those skills implies that they are experienced in multiple subject areas. A second fundamental reason lies in the domain-specificity of creativity and of critical thinking. Even though creativity and critical thinking can be discussed in a general way at the conceptual level, as if the domain of their application did not matter, in practice the fact that each requires knowledge and some level of expertise in a particular field means that they have to be practiced over and over in different fields. If they were domain-general, one could teach them in special creativity or critical thinking classes, or, for example, one could have the visual arts teacher be in charge of creativity, and the science or philosophy teacher be in charge of critical thinking.

But this is not the case. Creativity and critical thinking need to become a key objective of all subjects taught in schools, something that is reflected in the general perspectives across all subject areas in many Twenty-first Century school curricula. These general curriculum perspectives are commonly intended to allow students to develop some habits of mind, but the realisation of such intentions does require a mainstreaming of those learning objectives in all subject areas of education. While science teachers might feel that they are somewhat in charge of critical thinking, as science often challenges common wisdom, they could still emphasise more the remaining uncertainty of scientific "truths" (Rennie, 2020), even though the methods through which such "truths" are established are robust. Even more so, teachers should keep in mind that science requires creativity and imagination, as what may currently appear as the most obvious and conventional scientific statement was initially created by a very imaginative scientist.

References

Adler, L., Bayer, I., Peek-Brown, D., Lee, J., & Krajcik, J. (2017). *What controls my health.* https://www.oecd.org/education/What-Controls-My-Health.pdf

Davies, M. (2015). In R. Barnett (Ed.), *The Palgrave handbook of critical thinking in higher education.* Palgrave Macmillan.

Dennett, D. C. (2013). *Intuition pumps and other tools for thinking.* England: Penguin.

Ennis, R. (1996). *Critical thinking.* Upper Saddle River, NJ: Prentice-Hall.

Ennis, R. (2018). Critical thinking across the curriculum: A vision. *Topoi, 37*(1), 165–184. https://doi.org/10.1007/s11245-016-9401-4.

Facione, P.A. (1990). *Critical thinking: A statement of expert consensus for purposes of educational assessment and instruction.* Research findings and recommendations prepared for the Committee on Pre-College Philosophy of the American Philosophical Association. Retrieved from http://www.eric.ed.gov/ERICWebPortal/detail?accno=ED315423

Feynman, R. (1963). *The Feynman lectures on physics.* (Volume I: The New Millennium Edition: Mainly Mechanics, Radiation, and Heat.). Basic Books.

Feynman, R. (1955). The value of science. In R. Leighton (Ed.), *What do you care what other people think? Further adventures of a curious character* (pp. 240–257). Penguin Books.

Fullan, M., Quinn, J., & McEachen, J. (2018). *Deep learning: Engage the world, change the world.* Corwin Press and Ontario Principals' Council.

Guilford, J. P. (1950). Creativity. *American Psychologist, 5*(9), 444–454. https://doi.org/10.1037/h0063487.
Hitchcock, D. (2018). Critical thinking. In Zalta, E.N. (ed.), *The Stanford encyclopedia of philosophy* (Fall 2018 Edition). Retrieved from : https://plato.stanford.edu/archives/fall2018/entries/critical-thinking.
Kelley, T. (2001). *The art of innovation: Lessons in creativity from IDEO*. Currency: America's leading design firm.
Lubart, T. (2000). Models of the creative process: Past, present and future. *Creativity Research Journal, 13*(3–4), 295–308. https://doi.org/10.1207/S15326934CRJ1334_07.
Lucas, B., Claxton, G., & Spencer, E. (2013). Progression in student creativity in school: First steps towards new forms of formative assessments. In *OECD education working papers, 86*. Paris: OECD. https://doi.org/10.1787/5k4dp59msdwk-en.
Lucas, B., & Spencer, E. (2017). *Teaching creative thinking: Developing learners who generate ideas and can think critically*. England: Crown House Publishing.
McPeck, J. E. (1981). *Critical thinking and education*. New York: St. Martin's.
Mednick, S. A. (1962). The associative basis of the creative process. *Psychological Review, 69*(3), 220–232. https://doi.org/10.1037/h0048850.
Newton, L. D., & Newton, D. P. (2014). Creativity in 21st century education. *Prospects, 44*(4), 575–589. https://doi.org/10.1007/s11125-014-9322-1.
Paddock, W., Erwin, S., Bielik, T., & Krajcik, J. (2019). *Evaporative cooling*. Retrieved from : https://www.oecd.org/education/Evaporative-Cooling.pdf
Rennie, L. (2020). Communicating certainty and uncertainty in science in out-of-school contexts. In D. Corrigan, C. Buntting, A. Jones, & A. Fitzgerald (Eds.), *Values in science education: The shifting sands* (pp. 7–30). Cham, Switzerland: Springer.
Runco, M. A. (2009). Critical thinking. In M. A. Runco & S. R. Pritzker (Eds.), *Encyclopedia of creativity (pp. 449–452)*. Academic.
Schneider, B., Krajcik, J., Lavonen, J., & Samela-Aro, K. (2020). *Learning science: The value of crafting engagement in science environments*. United States: Yale University.
Sternberg, R. J., & Lubart, T. (1999). The concept of creativity: Prospects and paradigm. In R. J. Sternberg (Ed.), *Handbook of creativity* (pp. 3–14). England: Cambridge University.
Torrance, E. P. (1966). *Torrance tests of creative thinking: Norms. Technical manual research edition; Verbal Tests, Forms A and B, Figural Tests, Forms A and B*. Princeton, NJ: Personnel.
Torrance, E. P. (1970). *Encouraging creativity in the classroom*. United States: W.C. Brown.
Vincent-Lancrin, S., González-Sancho, C., Bouckaert, M., de Luca, F., Fernández-Barrerra, M., Jacotin, G., Urgel, J., & Vidal, Q. (2019). *Fostering students' creativity and critical thinking in education: What it means in school*. Paris: OECD. https://doi.org/10.1787/62212c37-en.

Chapter 4
Exploring STEM Learning in Primary Classrooms: In Support of Social Justice Agendas

Bronwen Cowie and Paula Mildenhall

Abstract This chapter explores how STEM education might be implemented with a social justice orientation in primary classrooms. The chapter initially explores how knowledge, empathy and action play a role in social justice-oriented STEM education that can support student capacity and inclination to take action for wider societal 'good'. As illustrated within the first of three vignettes from primary school classrooms in the chapter, knowledge is necessary but not sufficient; there is value in developing and exercising empathy and critical and creative thinking so that students are willing and able to take constructive action. Next, the chapter provides a vignette of how teachers can work to address the issue of equity of access, participation and achievement in STEM whilst also engaging with the community. Finally, the chapter moves beyond a focus on students' learning about a particular issue to considering how students might become change agents who are sources of authoritative knowledge relevant to their community. Overall, this chapter illuminates some of the challenges and opportunities of STEM education for primary school teachers and students when the focus is on how it can support a social justice agenda of equity of access and contribution to the common good.

Keywords STEM education · Social justice · Case study · Interdisciplinary

---

B. Cowie (✉)
School of Education, University of Waikato, Hamilton, New Zealand
e-mail: bcowie@waikato.ac.nz

P. Mildenhall
School of Education, Edith Cowan University, Joondalup, Australia
e-mail: p.mildenhall@ecu.edu.au

A. Berry et al. (eds.), *Education in the 21st Century*,
https://doi.org/10.1007/978-3-030-85300-6_4

4.1 Introduction

It is our contention that, while the focus on STEM education as contributing to economic competitiveness and employment is commendable, this focus provides only a limited vision of what STEM education might aspire to and achieve. In this chapter we set out three vignettes to illustrate what primary school STEM education can be if it were assumed that STEM education should support a social justice agenda of equity of access and contribution to the common good. We do not intend to imply that this should be the only aim of STEM education but rather that this aim has been significantly undervalued. Situating STEM education within a social justice oriented framework brings it in line with the wider aims of schooling that embrace a concern with developing student capabilities and competencies for active learning, including the central issues of creativity and critical thinking and contributing both lifelong and life-wide.

4.2 Establishing the Context

In this chapter we view STEM education as an interdisciplinary and applied approach whereby the four component disciplines are integrated "into a cohesive learning paradigm based on real-world applications" (Hom, 2014, p. 1). Ideally, STEM education not only provides access to science, technology, engineering and mathematics learning in some integrated form, but also fosters the capabilities needed for active citizenship and learning lifelong (Bybee, 2010; English, 2016; Partnership for 21st Century Skills, 2011; Fadel & Trilling, 2009). These capabilities include collaboration (the capacity to present, debate and negotiate ideas and tasks; OCED, 2005), creativity (thoughts and actions that lead to novel and appropriate production; Sternberg & Lubart, 1999) and critical thinking (questioning and perspective taking; Ellerton & Kelly, Chap. 2; Lancrin-Vincent, Chap. 3).

Recognition of the role STEM plays in national, societal and individual life opportunities and wellbeing means that differences in access to, participation and achievement in STEM education is a social justice issue. In response, policy and research is increasingly focused on "STEM for all" (e.g., Parker et al., 2016). This agenda aims to take account of the needs and resources that Indigenous, English as a Second Language, migrant and refugee students, girls and students from lower socioeconomic backgrounds bring to STEM education as a way of addressing these inequities. It also includes a concern about the challenges faced by disabled and special needs students (Basham & Marino, 2013; Moon et al., 2012), although on the whole, less attention has been paid to this group in relation to closing opportunity and achievement gaps. How students might be supported to understand and motivated to "advance the common good" (Nguyen & Walker, 2015, p. 243) through and with STEM ideas, dispositions and practices also demands attention from a social justice point of view. Zeyer and Dillon (2019) argue that both systemising and empathising are required to address complex situations such as those that involve 'Science|Environment|Health' issues. They define systemising as "the ability to perceive physical things and understand them and their function in the context of a system" (p. 297–298) and empathising as "an affective response to the directly perceived, imagined or inferred feeling state of another being" (p. 300). In their book, Leggon and Gaines (2017) illustrate that through appropriately designed programs students from all backgrounds can come to understand the contributions that they might make to STEM and to their community and society in general through STEM. In this chapter we are interested in how two aspects of social justice - access and contribution - play out in the context of STEM education primary classrooms.

The vision of equity of access or 'STEM for all' is often accompanied by a view of learning as spanning classroom and school boundaries (United States Department of Education, 2016; Madden et al., 2017; Penuel et al., 2016). Scholars pursuing this line of argument emphasise the potential for and value of mutual learning across schools, homes and communities, and between children/students and adults/family and community members. Students learning from family and community members and students sharing their learning to inform their families and community provide students with a goal for their learning that extends beyond the immediate classroom context (Chen & Cowie, 2013; Engle, 2006; Rennie, Chap. 7; Cheng & Leung, Chap. 8). Students utilising their school learning beyond the classroom is also a feature of social justice education. Social justice educators argue that knowledge and understanding are required for students to participate in "positive social change" (Hackman, 2005, p. 104). In addition, students need to know about strategies that can be used for social action and how to sensitively observe and consider the dynamics between different peoples, something Hackman suggests is supported by self-reflection. Hackman also points out that students need to be able to critically analyse what they know to "bring the power of that information to fruition" (p. 106). These aspects resonate with current goals for STEM education in their scope, suggesting a possible cross-fertilisation between the two. In the vignettes that follow, we aim to illustrate STEM education which adopts an integrated interdisciplinary approach, develops twenty-first century skills and also has a social justice orientation.

4.3 STEM Education in Action in the Primary Classroom

In this section we provide three vignettes of classroom practice to illustrate some of the ways that primary students can engage in STEM education via strategies that foster empathy, and critical and creative thinking, and that open up opportunities for engagement and action within the wider community.

4.3.1 Vignette 1: Students Developing Knowledge, Empathy and Action in STEM Learning

This first vignette illustrates how knowledge, empathy and action each play a role in STEM education when the goal is to foster student capacity and inclination to take action for wider common good. It is based on the module *The Long Walk* (see http://stemlearning.org.au/stem-learning-project), produced as part of the STEM Learning Project. This project developed STEM teaching modules aligned with the Australian Curriculum, including the General Capabilities (Australian Curriculum, Assessment and Reporting Authority, 2015). The STEM Learning Project modules aim to engage students in and challenge them to solve real-world problems of relevance to their communities by developing and applying conceptual understandings and engaging with the processes of the STEM disciplines in an integrated way. Creative and critical thinking, and collaboration, are explicit learning goals in all the project modules. The module design structure is underpinned by a four phase problem-based approach of research, investigate, design, and evaluate and communicate. The *Long Walk* module was evaluated as part of The STEM Learning Evaluation project (as were the other modules we refer to in this chapter). Evaluation data were collected via classroom observation, video, teacher and student interviews, and student work.

The *Long Walk* topic was developed to be of interest to generalist primary school teachers and their students because of media coverage on the plight of refugees. The module is anchored in the challenge of designing shoes for refugee children using only the resources that might be at hand in a refugee camp. In line with the module, the teacher involved in the evaluation study used photographs and whole class discussion to help children to connect with the refugee experience and the need for shoes (see Mildenhall et al., 2019a for full details). The students' empathetic response to viewing the photographs and participating in the discussion was summed up by the student who said: "I feel sorry for them because they didn't do anything to deserve to be poor and homeless because they did nothing wrong". The children listened to the "The Little Refugee" story authored by Anh Do. In the book Anh Do describes his experiences as a young boy travelling with his refugee family from Vietnam to Australia in 1980, and going to an Australian school.[1] Being a similar age and school level to the young Anh Do depicted in the book appeared to support the children to empathise with the day-to-day challenges Anh Do faced as a refugee. The teacher guided the children to use a Venn diagram to critically analyse and compare how the life of a refugee child was the same as and different to their own. The children were able to identify the abundant conditions of their lives in contrast to those of refugees. They were also able to identify how their lives were similar: "Yes, not all people will have mums and dads and everybody there but that's the

[1] As well as being a successful author of several books, today Ahn Do also has a high media profile as an actor, a comedian and an artist.

same as where we live, not everybody has a mum and a dad, but we both have families."

The children's appreciation of the conditions in a refugee camp, and hence of the functions shoes need to serve and the limitations on the resources that are available to produce shoes, was key to the development of the criteria for the shoes the students were to design. The students were able to conclude that the shoes needed to be waterproof, long lasting, comfortable and have a closed toe and a strong sole. These criteria then acted as product specifications and parameters (Australian Curriculum, Assessment and Reporting Authority, 2015). As a first step, the children conducted science investigations to explore the properties of a range of materials including sponge, drink bottles, rubber, and cardboard. They used the results to select the materials used for the different parts of their shoes. When the teacher asked the students how the design and final product were informed by the class investigations and their appreciation of the refugee situation, Tom[2] answered, "The rubber has good grip on the sole ... and it is long lasting because it is made of a very tough object". Sarah explained that her group had used rubber for the sole because it was waterproof. Each group produced a viable shoe. The investigations informed students' ability to critically analyse materials as a precursor to the need for them to be creative in the selection and use of very restricted resources to design and make a novel and appropriate pair of shoes (see Lancrin-Vincent, Chap. 3; also Kelly & Ellerton, Chap. 2, in that critical thinking is frequently a step in creative thinking and so the two are often employed together).

This vignette illustrates how STEM knowledge and STEM-based action can be developed through a social justice oriented context that resonates with children and evokes their empathy. As noted above, knowledge is necessary but not sufficient in this context. Students needed to develop and exercise empathy and critical and creative thinking for them to be both willing and able to take constructive action. As Zeyer and Dillon (2019) assert, both systemising and empathising were required to address the complexity of the refugee situation. Students were able to imagine the challenges refugee children faced, and acquire appropriate STEM knowledge and then employ their knowledge to act to meet the brief. Interestingly, empathy is the initial stage in the design thinking cycle proposed by Hasso-Plattner Institute of Design at Stanford (Dam & Siang, 2019) and features in design processes focused on user experience/human-centred design (Minichiello et al., 2018).

[2] All student and teacher names given in this chapter are pseudonyms

4.3.2 Vignette 2: Ensuring Access to STEM Education for Students with Special Capabilities

This vignette continues and extends our focus on knowledge building, empathy and action and their interaction by illustrating how teachers can work to address the issue of equity of access, participation and achievement in STEM. Specifically, it details how two teachers adapted and implemented the module *Every Bird needs a Home* so that it was appropriate for their Year 4 to 6 students. The students were enrolled in an education support centre[3] and had a range of complex abilities. For this module two teachers, Carol and Eve, worked together to adapt and teach the *Birds* module, which had been developed for Year 2 students (aged 7–8 years). The teachers selected the module because a raven with a broken wing known as Russell, had been visiting the school grounds for over two years and was of interest to their students. To ensure the module was 'personal and concrete' the teachers renamed it: *A home for Russell.* Carol explained, "We wanted to focus on something that the students would be really interested in rather than just talk about birds". They adapted the module for their students without compromising its focus on collaborative critical and creative thinking by arranging for the students to work in pairs rather than in small groups, and supported this by increasing adult support through the inclusion of Educational Assistants and modifying handouts to use more images and less written text in order to facilitate student responses. For example, during the first lesson the teachers focused the students on what birds would need to survive by using photographs of birds in different settings. The teachers' questioning scaffolded student thinking so they were able to reason about what birds needed to survive rather than simply recite facts they had been told. In the following dialogue Jack was able to explain why birds might have their nest up a tree and hence "in a safe place".

Teacher: Jack can you tell me something else that your group decided birds needed?
Jack: Big nest and laid eggs.
Teacher: Big nest, good. And where do they usually make their nest?
Jack: Up in the tree.
Teacher: Up in the tree. Why up in the tree?
Jack: So they won't be taken down.

The students then collected data on what birds were present (if any) at different sites around the school by photographing them and recording where they were on a map. The Educational Assistants and teachers helped each pair to tally the number of different birds on a template that included pictures of the different birds. The students then produced a pictograph as a literal graph. To further their understanding, Carol asked the students to represent their pictograph using unifix blocks and to explain to her what it meant. Unifix blocks are plastic linking cubes that snap

[3] In the education jurisdiction in which these vignettes occurred (the Australian state of Western Australia), "education support centres" provide additional services for students with special learning needs who also attend mainstream schools.

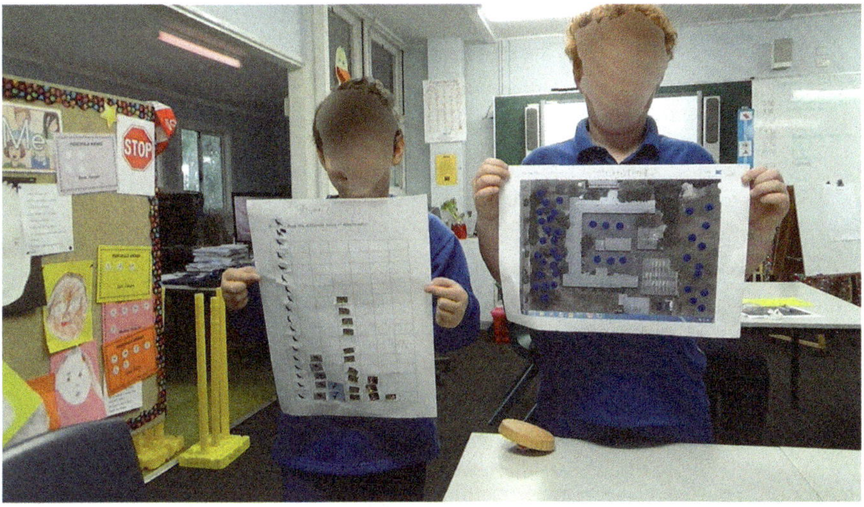

Fig. 4.1 Jack and Adam's completed map, pictograph

together and come in a variety of colours. Each pair was able to do this in some way. Figure 4.1 shows Jack and Adam's completed map and pictograph, and the unifix version of their pictograph.

In this example the teachers' adaptations of the standard unit were consistent with the tenets of the Universal Design for Learning (UDL) framework, which has been advanced as a way to develop a curriculum that all students can access (Moon et al., 2012; National Science Foundation, 2015). The UDL framework advocates for: (i) support for student engagement through the provision of choice and making the learning relevant to students' interests and goals (in this instance focusing the module on the needs of a bird in their immediate environment); (ii) multiple means of representation of information by the teacher (the inclusion of photographs etc.), and (iii) multiple means of expression by students (marking birds on a map, tallying, developing a pictograph and a physical representation of the pictograph).

The element of choice was introduced for the next step in the module, which required students to source birdhouse designs from the internet and combine them to design a birdhouse that would suit Russell. In selecting useful designs, students also needed to consider the constraints imposed by Russell's broken wing. The transcript below demonstrates that students were able to critically evaluate what features would meet this brief, and deliberately introduce an element of novelty, that is, they combined two designs and planned for two doors.

Jack: We're mixing up these two houses, we'll make two doors on each one.

Researcher: Jack what are the features you're taking from this one and what are the features you're taking from that one?

Jack: We're putting the entire frame from this one except for one thing, the one door. Maybe we're going to have a little idea and put maybe those two doors from the Maccas one and put it to the other American house.

Researcher: Right, do you think those doors are going to be big enough for Russell?
Jack: Mmm.... We're still unsure
....
Student: We can make it [the door] bigger than the picture.
Jack: I know but how are we going to do that if he has a broken wing? How is he going to fit in if he has a broken wing?
Teacher: How would you do that Jack, if he has a broken wing?
Jack: We could make like the shape of Russell at the back, a little picture of Russell and cut it out in wood and make a shape of him so he can just put in his broken wing.

For the final activity, as per the original module, the teachers invited a local community group (a 'Men's Shed'[4]) to assist in constructing a birdhouse. The Men's Shed leader was so impressed with the children's models that he made a working birdhouse that incorporated ideas from each of their models. He also commented on the value for the Men's Shed group in being involved: "It's a great project, as it connects our members with the broader community". He continued, "Projects with schools keep the members involved in the community and gives the guys' a sense of purpose". The teacher reported, "I think the benefit worked both ways. With my students, it gave them a sense of community, importance and the notion that their ideas mattered and would be listened to. ... And for the guys at the Men's Shed, it gave them a sense of purpose and helping others."

This vignette provides an example of how teachers can scaffold children with a range of abilities to achieve STEM goals to do with (i) science and mathematics (data collection, representation and communication) and (ii) technology and engineering (the development and enactment of design specifications) in a way that involves collaborative critical and creative thinking to analyse, synthesise and take account of constraints in a novel manner. It also illustrates how these students were able to translate their broader understanding into local action - designing a house for the injured raven that frequented their school grounds. Lastly, it demonstrates how community members can play an integral role in children's teaching and learning activities to the benefit of both the children and themselves. In this case the children did not have the specific skills or resources required to make a working birdhouse for Russell, but through the involvement of the Men's Shed, a functioning birdhouse, which is still in use two years later, was produced.

4.3.3 *Vignette 3: Students as Change Agents Who Are Sources of Authoritative Knowledge*

In this final vignette we move beyond a focus on students' learning about an issue to consider how students might become change agents who are sources of knowledge relevant to their community. The topic of our third vignette is the global decline in the number of honey bees (see Mildenhall et al., 2019b, for further details). Two

[4]The Men's Shed organisation supports local groups of men to come together to work on meaningful projects, see Australian Men's Shed Association, 2017

teachers (Abigail and Ray) combined their Year 4 classes (ages 9 and 10 years) for the module. The *Honey Bees* module began with a local apiarist visiting the class and explaining how she cared for her bees and their hives and the honey extraction process. Teacher Abigail explained that the visit had helped the children to understand how a beehive works:

> She really put it on a level so that the kids understand how a beehive works, between the queen bee and all the worker bees ... instead of from a video, they actually saw and smelt the bee wax and all the cones and the smoke machine and what she'd wear.

She further explained "I think for them it really got them out of the headspace that this is just a school thing. It's actually real life and it's having a huge impact on the world around us." That is, the apiarist talking about her work helped the children to see the plight of bees as an issue that was relevant to a wider 'community of interest' (Engle, 2006).

Next, the students researched which of the foods they eat were reliant upon pollination and the causes of declining honey bee numbers. They identified these causes as monocropping, the overuse of pesticides, urbanisation, foreign species and pollution. Through teacher scaffolding, the children were able to critically reflect on why these things might be contributing to the decline of the bees. For example, they deduced that pesticides were useful in protecting crops and, by thinking more deeply, they concluded that shoppers' desire for fruit and vegetables that look perfect means that bees could suffer through the overuse of pesticides. The students reasoned, "They [people] don't want insects to eat the crops so that they can eat it."

Together with the students, Abigail then established that, "We need to show people how important bees are". Teacher Ray negotiated with them that they would design a board game that explains to the community how important bees are and how to help bees. The class played a range of board games to help them develop the criteria for their board game. Subsequently, it was agreed that the games would: (i) follow a points system, (ii) take into account who and how many players there might be, and (iii) be fun to play. The following excerpt is from one group's description of their board game design:

> Bill: This is our design so far. So we've got some of the materials we're going to use. We're going to have some action cards so, like so, you can earn things and you use those things. And we've got a board. We're going to get a board and we're going to have to glue it together. And we're going to have some characters which we'll, we'll make them go around the board and so sections will have like monocrops and like insecticides. And we'll have a grassland section and .....
>
> Jesse: Urbanisation.
>
> Josh: And a fertilisation section.
>
> Bill: Urbanisation.

Below is the blueprint the group produced (Fig. 4.2). We can see on this blueprint the creation of action cards, counters, the use of a board and that the players have to travel around the board.

The children invited their families, other teachers and the wider community (including the apiarist and other beekeepers) to the school to play their games. Seventy-five people accepted the invitation. Video footage reveals that community

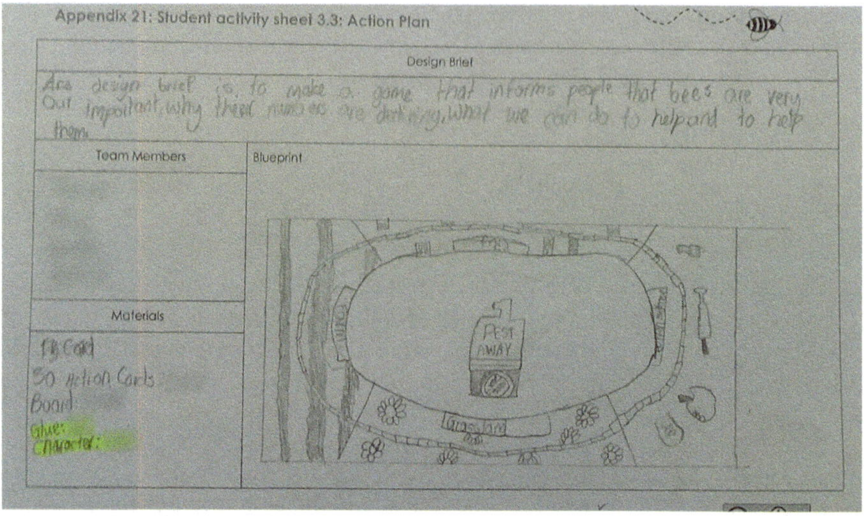

Fig. 4.2 The group's blueprint

members were enthusiastic about the games and the students were excited about running the games. Player responses indicated they found the games engaging and informative, as this comment suggests:

> I was saying how clever they were. It was amazing how much information they had and I was just incredibly impressed because some of the stuff I didn't know, actually. I was asking questions and they were able to answer everything. I didn't know what monopolinisation was and so they were able to tell me and were very articulate and very passionate, which was lovely to see. (Child's mother, as game player)

Reflecting on the success of the game day the children decided they wanted to take action to encourage bees and insects around their school. They decided to build insect sanctuaries. With the support of some parents the groups created bee-insect hotels, also known as Air Bee 'n' Bees (Green, 2015). These were hung in trees around the school grounds. When the three community members who volunteered were interviewed ten weeks after the game day, they stated the experience had inspired them to take practical actions such as planting more flowering trees and plants, and allowing vegetable plants, such as broccoli, to flower.

This vignette illustrates that STEM education can, and we argue should, support mutual and reciprocal learning between school and home/the school community - between students and their families and community members. Community members were captivated by the children's knowledge and their passion to improve the plight of bees, or in Zeper and Dillon's (2019) terms, their systemising and empathising about the plight of bees. Also of interest, the students initiated action at the immediate local level of their school grounds. This pattern of taking action locally and beyond teacher planned activities is something we have noticed in other studies. In one example, students followed up with their families to explore and take action to address issues related to the danger faced by native birds in New Zealand (Chen

& Cowie, 2013). In our view these longer timeframe actions are significant because they not only expand the timeframe for STEM learning, they provide evidence of students' willingness and capacity to embrace creativity and critical thinking as they exercise agency as authoritative contributors to their community.

4.4 Reflective Comments

Through this chapter we sought to illustrate some of the ways teachers might practice a STEM pedagogy that is attentive both to knowledge building and to how their students might, as socially and civically responsible citizens in the present and future, influence societal culture and action in the direction of social justice. Obviously, this is a task that extends well beyond a developmental project such as the STEM Learning Project, and well beyond what can be addressed in one book chapter, but we hope we have illustrated that STEM education can be multidimensionally valuable. Specifically, we have aimed to illustrate the potential outcomes of STEM education when it is focused on authentic 'real world' issues through an integrated approach which can support student learning in the different STEM disciplines. In our vignettes, and as Krajcik and Delen (2017) also found, it appeared that the need to produce a physical artefact provided direction and motivation for children's critical and creative thinking. The topic and outcome task for each of our vignettes included a social justice aspect - understanding the life and needs of refugee children prompted empathy and motivation for shoe design in the Long Walk; the needs of a bird with a broken wing provided a focus for the design and production of a birdhouse; and the implications for all countries and peoples of the decline in honey bee numbers provided the impetus for action in the final vignette. In each case students were supported to develop STEM content knowledge relevant to a real-world issue, to conduct an investigation related to possible solutions and then to design and test a possible solution. The development of knowledge was entangled with the students developing empathy or commitment to those involved (refugee children, Russell with his broken wing, and the honey bees). In making this point we return to our earlier analysis of the role of knowledge and empathy as important considerations in technological design and in understanding and addressing complex issues (Zeper & Dillon, 2019). Hence knowledge, critical and creative thinking and empathy are vital for both students and teachers within STEM-based learning when social justice understandings and the 'common good' are possible and desired outcomes.

Children sharing the STEM knowledge and insights developed in the classroom with their families and the community has received only limited research attention (Rennie, Chap. 7, this volume, provides a clear exception). The Birds and Honey Bees vignettes provide small scale examples of how students and teachers connecting with their school community can support mutual learning and be an effective strategy for multiplying the impact of school programs beyond the boundaries of the classroom (Falk et al., 2015). This extension is important because, as Ballantyne,

Connell and Fien (1998) point out, it is not enough to activate children as agents of change when many of today's challenges require timely action. It is adults who have the power to influence policy and practice in the short-to-medium term. The focus on adult learning also has the benefit of signalling to students the possibilities and value of STEM learning lifelong and life-wide, an important curriculum agenda throughout the world. This learning aspect recognises that there can be mutual and intergenerational benefit when teachers engage students and community members in the teaching and learning process. Making this connection moves us beyond enriching the curriculum to a focus on assisting students to see the wider relevance of what they are learning (Sias et al., 2017).

There are clear overlaps between STEM and social justice education in the concern to create equitable opportunities for all students. There are also significant differences (Sondel et al., 2017), namely that STEM approaches to equity tend to focus on the creation of pathways for individuals to gain access to the economic and social mobility benefits of STEM, whereas social justice education is more concerned with "preparing citizens with the capacities and commitments to interrogate and rearrange the very structures that maintain a stratified society in service of the common good" (p. 40). If STEM education is as successful as educators hope in creating a workforce that can both innovate and enhance economic wealth *and* social wellbeing then STEM workers, as influential citizens, will need to have developed the understandings, capabilities and motivation to use their STEM knowledge and skills wisely so that tomorrow's society is both equitable and just.

Acknowledgements We would like to acknowledge the funding for the research project that was provided by the STEM Education Consortium (Western Australia). We also thank and acknowledge the teachers and students who contributed data to the projects described in this chapter.

References

Australian Curriculum, Assessment and Reporting Authority. (2015). *General capabilities Australian Curriculum*. Retrieved from https://www.australiancurriculum.edu.au/f-10-curriculum/general-capabilities

Australian Men's Shed Association. (2017). *What is AMSA?* Retrieved from https://mensshed.org/about-amsa/what-is-amsa/

Ballantyne, R., Connell, S., & Fien, J. (1998). Students as catalysts of environmental change: A framework for researching intergenerational influence through environmental education. *Environmental Education Research, 12*(3–4), 285–298.

Basham, J. D., & Marino, M. T. (2013). Understanding STEM education and supporting students through universal design for learning. *Teaching Exceptional Children, 45*(4), 8–15.

Bybee, R. (2010). Advancing STEM education: A 2020 vision. *Technology and Engineering Teacher, 70*(1), 30–35.

Chen, J., & Cowie, B. (2013). Engaging primary students in learning about New Zealand birds: A socially relevant context. *International Journal of Science Education, 35*(8), 1344–1366.

Dam, R., & Siang, T. (2019). *5 stages in the design thinking process*. Retrieved from https://www.interaction-design.org/literature/article/5-stages-in-the-design-thinking-process

Engle, R. (2006). Framing interactions to foster generative learning: A situative explanation of transfer in a community of learners classroom. *The Journal of the Learning Sciences, 15*(4), 451–498.

English, L. (2016). STEM education K–12: Perspectives on integration. *International Journal of STEM Education, 3*(3), 1–8.

English, L. (2017). Advancing elementary and middle school STEM education. *International Journal of Science and Mathematics Education, 15*, 5–24. https://doi.org/10.1007/s10763-017-9802-x.

Fadel, C., & Trilling, B. (2009). *21st century skills: Learning for life in our times*. San Francisco, CA: Jossey-Bass.

Falk, J., Dierking, L., Osborne, J., Wenger, M., Dawson, E., & Wong, B. (2015). Analyzing science education in the United Kingdom: Taking a system-wide approach. *Science Education, 99*(1), 145–173. https://doi.org/10.1002/sce.21140.

Green, A. (2015). *Air Bee'n'Bee: Have backpack, will travel*. Retrieved from https://blog.csiro.au/air-beenbee-have-backpack-will-travel/

Hackman, H. (2005). Five essential components for social justice education. *Equity & Excellence in Education, 38*, 103–109. https://doi.org/10.1080/10665680590935034.

Hannon, V., Patton, A., & A., & Temperley, J. (2011). *Developing an innovation ecosystem for education*. San Jose, CA: Cisco Systems.

Hom, E. J. (2014). What is STEM education? Livescience. Retrieved from https://www.livescience.com/43296-what-is-stem-education.html.

Krajcik, J., & Delen, I. (2017). Engaging learners in STEM education. *Estonian Journal of Education, 5*(1), 35–58. https://doi.org/10.12697/eha.2017.5.1.02b.

Leggon, C., & Gaines, M. (2017). Introduction and overview: STEM and social justice: Teaching and learning in diverse settings. In C. Leggon & M. Gaines (Eds.), *STEM and social justice: Teaching and learning in diverse settings* (pp. 1–6). Cham, Switzerland: Springer International.

Madden, P., Wong, C., Cruz, A., Olle, C., & Barnett, M. (2017). Social justice driven STEM learning (STEMj): A curricular framework for teaching STEM in a social justice driven, urban, college access program. *Catalysts, 7*(1), 24–37.

Mildenhall, P., Cowie, B., & Sherriff, B. (2019a). A STEM extended learning project to raise awareness of social justice in a year 3 primary classroom. *International Journal of Science Education, 41*(4), 471–489.

Mildenhall, P., Sherriff, B., & Cowie, B. (2019b). The honey bees game: Engaging and inspiring the community with STEM. *Research in Science & Technology Education, 39*(2), 225–244. https://doi.org/10.1080/02635143.2019.1687440.

Minichiello, A., Hood, J., & Harkness, D. (2018). Bringing user experience design to bear on STEM education: A narrative literature review. *Journal for STEM Education Research, 1*, 7–33. https://doi.org/10.1007/s41979-018-0005-3.

Moon, N., Todd, R., Morton, D., & Ivey, E. (2012). *Accommodating students with disabilities in science, technology, engineering, and mathematics (STEM)*. Atlanta, GA: Center for Assistive Technology and Environmental Access, Georgia Institute of Technology.

Nadelson, L., & Seifert, A. (2017). Integrated STEM defined: Contexts, challenges, and the future. *The Journal of Educational Research, 110*(3), 221–223. https://doi.org/10.1080/00220671.2017.1289775.

National Science Foundation, National Center for Science and Engineering Statistics. (2015). *Women, minorities, and persons with disabilities in science and engineering: 2015*. Arlington, VA: Author.

Nguyen, T., & Walker, M. (2015). 'Capabilities-friendly' assessment for quality learning. *South African Journal of Higher Education, 29*(4), 243–259.

OECD, D. (2005). *The definition and selection of key competencies*. Retrieved from https://www.oecd.org/pisa/35070367.pdf

Parker, C., Pillai, S., & Roschelle, J. (2016). *Next generation STEM learning for all: A report from the NSF supported forum*. Waltham, MA: Education Development Centre.

Partnership for 21st Century Skills. (2011). *Framework for 21st century learning 2-page*. Retrieved from http://www.p21.org/our-work/p21-framework.

Penuel, W. R., Clark, T. L., & Bevan, B. (2016). Infrastructures to support equitable STEM learning across settings. *Afterschool Matters, 24*, 12–20.

Sias, C., Nadelson, L., Juth, S., & Seifert, A. (2017). The best laid plans: Educational innovation in elementary teacher generated integrated STEM lesson plans. *The Journal of Educational Research, 110*(3), 227–238. https://doi.org/10.1080/00220671.2016.1253539.

Sondel, B., Koch, J., Carrier, S., & Walkowiiak, T. (2017). Toward a theory of teacher education for justice oriented STEM. *Catalysts, 7*(1), 38–52.

Sternberg, R. J., & Lubart, T. I. (1999). The concept of creativity: Prospects and paradigms. In R. J. Sternberg (Ed.), *Handbook of creativity* (pp. 3–15). Cambridge: Cambridge University.

Traphagen, K., & Traill, S. (2014). *How cross-sector collaborations are advancing STEM learning*. Retrieved from http://www.noycefdn.org/documents/STEM_ECOSYSTEMS_REPORT_140128.pdf

U.S. Department of Education. (2016). *STEM 2026: A vision for innovation in STEM education*. Retrieved from https://www.air.org/resource/stem-2026-vision-innovation-stem-education.

Zeyer, A., & Dillon, J. (2019). The role of empathy for learning in complex science|environment|health contexts. *International Journal of Science Education, 41*(3), 297–315. https://doi.org/10.1080/09500693.2018.1549371.

Chapter 5
There's Something About James

Cathy Buntting and Alister Jones

Abstract Using a case study of teacher James and his class of 12–13 year olds building simple hydraulics machines, this chapter demonstrates the range of knowledge and skills needed to scaffold students' STEM learning and the value of focused learning conversations to support students' creativity and critical thinking. In particular, the chapter demonstrates James' ability to expertly navigate scientific, mathematical, technological and everyday discourses, knowing which discourse to use when and where in order to support students' conceptual and skill development throughout the unit of work.

Keywords Senior primary/elementary · Learning conversations · Discourse · Teacher roles · Case study

5.1 Introduction

The recent international focus on integrated STEM education in many ways echoes earlier calls for enhancing science education by using meaningful contexts for learning. For example, educational calls through the 1970s, 80s and 90s to teach science in context, as well as the science-technology-society (STS) and science-technology-society-environment (STSE) movements, can be interpreted as earlier steps towards identifying cross-curricular opportunities that emphasise the relevance and usefulness of science for everyday life. In other words, integrated approaches to science education are not new. However, the widespread more recent focus on STEM has led to a plethora of new resources, activities, kits and websites, and the galvanising of a range of curriculum discussions at multiple levels. For example, conversations about STEM at a macro (policy) level are raising the profile of the STEM disciplines in school curricula across many educational jurisdictions. At a meso (school) level, conversations about STEM are similarly prompting school leaders and teachers to consider the role and prominence of STEM teaching and

C. Buntting (✉) · A. Jones
University of Waikato, Hamilton, New Zealand
e-mail: cathy.buntting@waikato.ac.nz; alister.jones@waikato.ac.nz

A. Berry et al. (eds.), *Education in the 21st Century*,
https://doi.org/10.1007/978-3-030-85300-6_5

learning in the whole-school curriculum (at primary school), and opportunities for curriculum integration (at secondary school). At a micro (classroom) level, we suggest that it is the conversations in classrooms that are key to scaffolding students' conceptual and procedural learning across the STEM disciplines, as well as fostering positive attitudes towards STEM.

As Amanda Berry points out in Chap. 11, this volume, the dominating policy-level arguments put forward for STEM education relate to the economic and social progress of nations – the need for a stronger pipeline into STEM careers, and for an educated citizenship who can engage meaningfully with the many 'wicked problems' facing societies world-wide. STEM education is also identified as a means for developing high-value competencies like creativity, critical thinking and collaboration, and is heralded as a vehicle for engaging students in relevant, future-focused school learning.

In contrast, voices calling for a more considered response to STEM fever (or Berry's 'GERM' phenomenon, see Chap. 11, this volume) have focused on the potential for disciplinary-specific knowledge and skills to get 'lost' in the integrated learning contexts created by a STEM approach. Others point to the challenges associated with teacher expertise, and the dominating influences of the teacher's subject sub-culture in creating a STEM curriculum. For example, it is not unreasonable to expect that a science educator is more likely to create STEM programmes that give prominence to science learning outcomes, whereas a technology or mathematics educator might prioritise predictably different learning outcomes.

As both educators and researchers with expertise that straddles both science and technology education, we are sympathetic to concerns that certain disciplinary foci might be prioritised, or equally that the unique nature of the different disciplines might be underplayed. However, we have also long argued for the use of technological examples to richly contextualise and empower science learning (see short reviews in Jones, 2009; Jones & Buntting, 2015), and we see the STEM movement as giving voice to this approach. Our pragmatic position is that the current fascination that many policy makers, school leaders and teachers have with STEM provides a fertile context for conversations about contemporary priorities for school teaching and learning, including in science and technology.

In thinking about STEM education and its potential for supporting creativity and critical thinking, we are also acutely aware of the wide-ranging knowledge and skills demonstrated by teachers who effectively integrate science and technology education, let alone STEM. To demonstrate the expertise required, this chapter presents an in-depth case study of a STEM teacher, James. Drawing on a classroom unit in which 12–13 year olds built simple hydraulics machines, we demonstrate James' ability to effectively facilitate targeted learning conversations to ensure the students achieved multiple conceptual and procedural learning outcomes. The purpose is to showcase the knowledge and skills that are required to initiate and guide such conversations as expertly as does James, and the many roles that effective STEM teachers are likely to exemplify.

5.2 Learning Conversations

The role of focused conversations in supporting learning is well established, as evidenced in an extensive volume of literature in a wide range of educational contexts (e.g., Black, 2013; Black & Harrison, 2004; Moreland et al., 2008; Moreland et al., 2009). The importance of learning conversations is also a key foundation of both social constructivist and sociocultural learning theories – where interactions with others is understood to support learning. Indeed, Shulman (1987), in his seminal work on teacher expertise, drew attention to the skills associated with the "management of *ideas* within classroom discourse" (p. 1, emphasis in original). He went on to argue that analysis of the management of ideas is as important as analysis of classroom management practices "if our portrayals of good practice are to serve as sufficient guides to the design of better education" (p. 1).

The research presented in this chapter focuses on the pedagogical practices of James, and how these manifested in the conversations that he had with students and the ways in which he supported students to talk, work and learn together. To focus our discussion, we draw on earlier thinking in which we identified attributes for effectively using relevant, authentic, meaningful contexts in science education (Corrigan et al., 2012; see Fig. 5.1). These attributes include **fluency** (the ability to move seamlessly between the context being used, and the scientific concepts and processes that students are intended to learn), **disposition** (the inclination, or desire, coupled with the ability to effectively use context-based approaches), **discernment** (shrewd decision making that requires both fluency and disposition), and **competence** (the mastery of fluency, disposition and discernment). Our intention in this chapter is to explore what 'competence' might look like in an integrated unit that includes both science and technology.

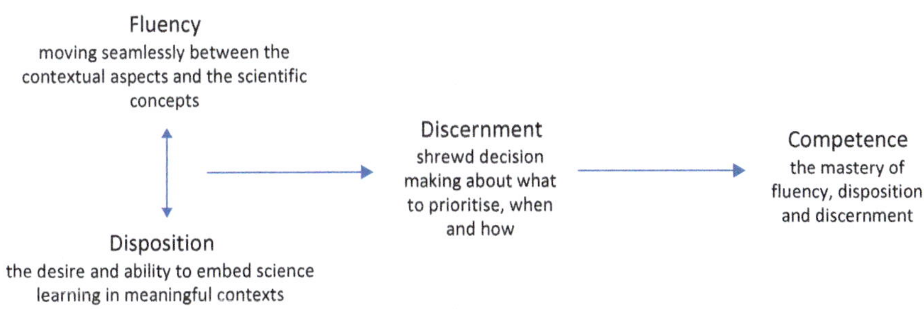

Fig. 5.1 Teacher attributes for using relevant, authentic contexts to deepen science education. (Corrigan et al., 2012)

5.3 The Example of a Hydraulics Unit

To explore what pedagogical ‘competence’ might look like in a primary-level STEM unit, we collaborated with James, an experienced teacher recognised within his local community for his innovative and future-focused teaching practices, and who was a STEM education leader in his school. James was extremely receptive to the opportunity to contribute to the research project, and provided a warm welcome to his classroom. An interpretive approach guided the research, which focused on understanding the nature of the learning conversations that took place in the classroom, and how these supported the students’ learning. The first author (Cathy) adopted the role of observer-participant for the unit’s seven full-day classroom sessions held on consecutive Mondays during the second school term. The research data included field notes, audio-recordings of discussions (between James and the students, between small groups of students, between the researcher and James, and between the researcher and students), photographs taken by the students and researcher, and reflective videos created by the students.

The study took place in New Zealand, where the national curriculum provides a broad set of learning aims and schools have autonomy to develop a curriculum that is locally relevant. The overall framework of the curriculum identifies ‘values’ and ‘key competencies’ to be embedded across all curriculum planning, and ‘science’ and ‘technology’ are specified as two of the eight mandated learning areas (Ministry of Education, 2007). There are no specific learning outcomes for engineering, although the definition of technology is suitably encompassing of a general definition for engineering:

> Technology is intervention by design: the use of practical and intellectual resources to develop products and systems (technological outcomes) that expand human possibilities [...] Quality outcomes result from thinking and practices that are informed, critical, and creative. (Ministry of Education, 2007, p. 32)

Science, by contrast, is specified as

> a way of investigating, understanding, and explaining our natural, physical world and wider universe. It involves generating and testing ideas and gathering evidence [...] in order to develop scientific learning, understanding, and explanations. Scientific progress comes from logical, systematic work and from creative insight, built on a foundation of respect for evidence. (Ministry of Education, 2007, p. 28)

As can be seen, creativity and critical thinking are common to both these descriptions, although the *purposes* for these in each learning area context are different. In addition, schools are specifically encouraged to “make use of the natural connections that exist between learning areas and that link learning areas to the values and key competencies” (Ministry of Education, 2007, p. 16). Values include innovation, inquiry and curiosity. There are five key competencies: thinking; using language, symbols, and texts; managing self; relating to others; and participating and contributing.

5.3.1 The Classroom Context

At the time of the research, James had been teaching in New Zealand for 24 years in schools across the socioeconomic and rural-urban spectrum. He considered his main strength to have been physical education and sport, until 5 years previously when he "took on IT" at his previous school. He had been at his current school for 4 years. Having introduced STEM as a classroom teacher 3 years previously, he was subsequently supported by the principal to encourage a school-wide STEM curriculum. He indicated that his STEM work arose as a result of personal interest – he had grown increasingly aware of educational conversations about STEM/STEAM, the focus in STEM/STEAM on inquiry learning and on group collaboration appealed to him, and he had a group of hitherto disengaged boys in his class. In 2017, the release of a new national digital technologies curriculum focused the direction of students' digital learning across New Zealand, and at James' school it meant that he moved into a newly-created role of Digital and STEAM Leader. In practice this means that he now spends time supporting teacher colleagues through team-teaching and providing ongoing follow-up support.

In the senior school, Year 7–8 (12–13 year olds) students had previously accessed a technology programme at a local secondary school. However, transport to that school had been time-consuming and problematic, and in the year prior to this research James' school had developed and implemented its own programme – a choice of STEM-related options that students could select from and participate in for a 10-week school term. Each of these options had a strong technology underpinning. The programme was labelled the 'IGNITE' programme, an acronym developed around the school's interpretation of key elements of the technological process:

Ignite our thinking and imagination (identify stakeholders and design a brief)
Gather our questions (carry out research)
New information is collected (develop conceptual designs)
Into the learning pit (prototype development)
Tying our learning together (final design and manufacture)
Evaluate and celebrate our learning (evaluation).

Posters outlining this process were displayed in each of the five IGNITE classroom spaces, and each IGNITE option used a shared but customisable online template in which students curated artefacts showcasing their learning.

During the period of this research, five IGNITE options were available, of which students could choose one: Construct (hydraulics), Conduct (electricity), React (chemistry), Create (fabrics), and Explore (constructing a culturally-appropriate wooden entrance way and garden). This was the first time that James had offered the hydraulics option, which he developed having come across some simple hydraulics kits online. As he explained:

> I saw that someone had made a hydraulic arm out of cardboard. What struck me was the cardboard side of it – I thought, this is quite a good STEM activity. I looked into it a bit more, and you could buy kits online. I thought this would work quite nicely with IGNITE. It has the construct side, but it also has the science aspect.

Eighteen students selected the hydraulics option – 12 boys and six girls (an extra teacher is employed for the IGNITE programme to keep class sizes small). Half the students were familiar with the IGNITE format, including the school's technological process, having participated in the programme the previous year.

5.3.2 The Science of the Hydraulics Unit

The science of hydraulics was introduced part way through the first day of the hydraulics IGNITE programme. James was working through the students' online workbook and had a page titled 'Technology and science go hand in hand.' This page listed the Ministry of Education's (2014) five science capabilities: gather and interpret data, use evidence, critique evidence, interpret representations, and engage with science (defined as "using the other capabilities to engage in 'real life' contexts"). Using whole-class discussion, James explored ways in which each of these capabilities related to the proposed project, i.e., to build a simple hydraulics machine.

This introduction led seamlessly into a discussion, "What are hydraulics?" In an opening brainstorm, students' ideas about hydraulics were initially vague: words that were first suggested included water, science, and electricity. One student then provided a more complete definition that James wrote down on the board: "It's a system that uses pressure and water to move things." Another student added, "It uses force," to which a third countered, "Pressure is force." Others were able to point out that a hydraulics system doesn't necessarily use water, it could use oil or other liquid. In other words, by pulling out the ideas represented across the class, James worked with the students to co-construct a useful working definition. Students then used their digital devices to find out more about hydraulics, accessing three YouTube clips that James shared with them. These videos collectively highlighted three important concepts: liquids are virtually incompressible, pressure is transmitted evenly through a liquid system, and a hydraulic system is a force multiplier (the force on one piston can produce a larger force on a larger piston). James wrote two questions on the board to guide students' engagement with the videos: (1) What is Pascal's Law? and (2) What is the difference between hydraulics and pneumatics? The students each used their own devices to watch these clips. James explained that this was so that they could pause and rewind where they needed to. He circulated around the class while they were doing this, asking students about what they were discovering, then re-convened a whole-class discussion during which students were required to have their devices closed.

After discussing what the students had learned, James set the scene for some scientific fair testing. First, the students discussed as a whole class what a 'fair test' is; they then brainstormed different variables that could be tested through a simple hydraulic system (two syringes and tubing). Working in self-selected pairs or small groups, the students spent the rest of the day learning through play. To support this, they had access to syringes of different sizes, lots of tubing that could be cut to different lengths, and water, oil and treacle ('golden syrup') to test the impact of viscosity. To help scaffold their explorations, students set up google sheets in which to record the research question, equipment, method, results and conclusions of their investigations.

James circulated around the class throughout the session, actively and deliberately engaging the students in focused learning discussions. Often these related to the idea of a 'fair test'. For example, one pair of students were testing different lengths of tubing, but had set up each of three systems by filling one of the syringes in each system with water. As a result, across the systems, there was more air in the system with the longest tubing. "Is this a fair test?" James probed. This led to a discussion about the possible impacts of air in the hydraulic system, and cued James to check in with each group that they were 'bleeding' the hydraulic system to remove all the air (except for cases where students were deliberately testing the impact of having air in the system). The high level of engagement by all students was noticeable, with students coming in after their lunch break eager to continue working on their fair tests. In a concluding whole-class discussion near the end of the school day, James asked the students to reflect on what they'd learned. One student's summing up was met with much other nodding: "It's been so fun! We've learnt lots of new things. I never really knew there were so many things about hydraulics."

Day 2 of the 7 day unit was again committed to scientific thinking and fair testing. James, having reflected on the students' work in the previous session, recognised that although all students had been deeply engaged, several had not been able to hone their exploratory investigations into a scientific test. To begin the lesson, James facilitated a whole-class discussion to remind the students about some of the key concepts underpinning hydraulics systems. When asked about Pascal's law, one student was able to easily articulate that "All force in a liquid is transmitted equally throughout the liquid," and James wrote this onto the board. After this discussion, the students went on a 'hydraulics hunt' around the school, taking digital photos of hydraulics systems using their devices. James was outside with them, interacting with the different groups and sending them after one another when a particular group located an example that he wanted them all to see. Back in class, various photos were shared with the whole class via *AirPlay* to a large shared screen. James expertly guided discussion about what is, what is not, and what could be a hydraulic system. For example, he used a syringe to demonstrate the working of the hydraulic suspension of the large truck trailer that was temporarily parked on the school grounds. He explained how hydraulics operate car and bike brake systems. He asked one student to google and then explain how hydraulics are used in gear changing in some bikes. He used probing questions to remind students that the key feature of hydraulics is the movement of liquid to generate an action – why the opening and closing of a tap is not a hydraulic system, for example, and why the high-level window opening mechanism in the school gym may or may not be hydraulic. James was able to comfortably guide this conversation because of both his prior reading about hydraulics, and his general knowledge about simple machine mechanisms.

James used this discussion to set up the next activity, which was to revisit the fair testing from the previous week. He pointed out to the class that, "A lot of what you did was observing and trying stuff out." This time, to focus the activity, the students were required to select a specific variable to test and set their test up as a display (cardboard and pipe cleaners were provided; the pipe cleaners were used to secure the syringes to the cardboard). The students were also asked to create a short video explaining their fair test and findings, and link their displays to their videos by generating a QR code. These displays were then made available in the school staffroom. In other words, the scientific investigation was framed as a technological process, with the brief: "Create a static display and short explanatory video outlining your results."

After a quick reminder brainstorm about the variables that could be selected, James used the example of testing different syringe sizes to highlight the importance of keeping all other variables constant (e.g., the length of tubing in each system), including bleeding the system. While working in pairs or small groups, several students revisited the idea of bleeding the system with James, asking why it matters. Here, James referred back to both an earlier observation some students had made, that air can be compressed but liquids can't, and Pascal's law, that the force must be distributed evenly through the system and that air interferes with this.

Close observation of one student working on his own demonstrated a level of criticality about what he was observing. This student had elected to test the effect of different syringe size. However, he noticed that the patterns in his findings were inconsistent. He explained:

> It's strange. We need to figure out how it's working. The thing is, this [medium syringe] to this [large syringe] is quite easy [when you push it]. This [large syringe] to this [medium syringe] is hard. This [medium syringe] to this [medium syringe] and this [medium syringe] to this [medium syringe; working in the opposite direction] is both easy. This [medium syringe] to this [small syringe] and this [small syringe] to [medium syringe] is both medium. It's just so confusing. We need to try and figure out how this is working. It's being very weird.

Discussing the lack of an apparent pattern with James, this student came to the conclusion that some of the syringes had been used the previous week and were still sticky (having been used with the syrup), and therefore difficult to push regardless of the system they were in; this student's critical thinking highlighted for the class the need to be using clean syringes, or another variable might inadvertently be introduced. James subsequently checked that this was the case with other student groups.

By the end of the school day, a range of displays had been created, along with students' explanatory videos. As in the previous week, engagement remained high throughout the day – largely, it seemed, because the students responded well to the autonomous environment created by James.

5.3.3 The Technology of the Hydraulics Unit

While the first part of day 1 had been committed to an introductory exploration of what technology is, the focus of the learning then moved onto a scientific exploration of hydraulic systems. On day 3, the technology angle shifted back into the spotlight. Again, James started the lesson with a whole-class reminder brainstorm about what had been learned in the previous 2 days. He then moved on to talking about the technology cycle. As previously indicated, for the IGNITE programmes this included: identify stakeholders and design a brief; carry out research; develop conceptual designs; prototype development; final design and manufacture; and evaluation. In this instance, an external stakeholder was not identified, and the brief was very general: 'Build a simple machine with at least one hydraulic system'. To get started, the students were tasked with brainstorming ideas for their machines, again in self-selected pairs or small groups. One student had also brought in a bottle of commercial hydraulic fluid that James talked about with the class. This unexpected interaction helped to reinforce the pervasiveness of hydraulic systems in our everyday lives – something that James had sought to show through the 'hydraulics hunt' on Day 2.

While students discussed ideas for their own hydraulics machines and looked for inspiration from YouTube, James circulated around the class and asked the groups what they were looking at, what the machines could do, and where the hydraulics were. James brought the class together to talk about some of the ideas that had been discussed: scissor lifts, digger arms, and a hydraulics-controlled steady-hand maze. James noted that many of the ideas included more than one hydraulics system, and that students would need to have a way of working out which syringe controlled which movement. One student group suggested using food colouring, which they'd seen in action in a YouTube clip they had watched; this idea was subsequently adopted across the projects. Students were then encouraged to start working on plans for their machines. Here, James relied on the fact that some of the students had worked on plans as part of their IGNITE work the previous year, discussing and clarifying the process and purpose of drawn plans with student groups rather than with the whole class. At various points he referred back to the scientific fair testing that had been done, for example, asking whether it would matter how long the piece of tubing was that linked the two syringes. By way of supporting one group to work together more collaboratively, he encouraged them to "Treat the project like a work site", identifying different roles and who was responsible for what. In another discussion, James exhorted a group to be specific in their planning: "Do you see the builder standing with the architect? No, generally the builder works just from the plans. That's how good the plans need to be!" While all the groups drew on YouTube videos for inspiration, most of which showed some of the steps taken to build the machines, none of these videos showed specific designs.

By lunch time planning for the following projects was well underway: three digger arms, two scissor lifts, one maze, one crusher, and one toy retrieval arcade game. Most students were in groups of 2–4, although one boy elected to build a maze on his own and another boy drifted between groups in this and the subsequent sessions. Demonstrating James' responsivity to students' individual needs, on day 5 he gave this boy a hydraulic kitset to work on in order to help him to focus.

In relation to the planning, James focused his discussions with the groups on the importance of scale, keeping measurements simple (using whole centimetres), showing how the syringes would be attached, and showing the different parts that would be needed. The building of prototypes began after lunch, with a range of materials and equipment available: syringes of different sizes, tubing, wooden toothpicks and skewers, small rubber tubing to secure the skewers, popsicle sticks (for the scissor lifts), food colouring, cardboard, rulers, scissors, glue guns, and a compass for drawing circles.

Days 4 to 7 of the unit were committed to constructing prototypes, primarily using cardboard. Students focused on constructing, critically evaluating the different steps involved, identifying and solving problems as they arose, updating their plans, and then constructing a final product, usually using MDF. Across these sessions, additional materials and equipment were also available: MDF, thicker wood, thick wire, wood glue, power drills, screws and screwdrivers, sanders, safety goggles, a handsaw and a scroll saw. James also sourced additional materials as needed, for example, clear plastic needed for the toy retrieval arcade game.

James actively engaged in purposeful discussions with students throughout the construction phase; students would approach him directly wherever he was in the class, and he would also target his engagement by approaching a group and inquiring about what they were doing. In addition, he was constantly monitoring the environment to ensure that students' practice was safe, explaining or demonstrating how to use the equipment when necessary, and encouraging everyone in a group to have a go, for example, at drilling or sawing. Several of the students had used drills, saws and sanders in the previous school year, and some had used them at home; for others this was their first time. James also helped the students to focus on working accurately and carefully – using rulers to draw straight lines ("Is that supposed to be a square?"); measuring three times, cutting once; checking that components were working as expected; and in the case of the scissor lifts, carefully aligning the different parts to reduce tilting. The importance of plans and prototypes was revisited multiple times, and James persistently exhorted students to think critically about what they were doing, what was working well in their machines, the problems they were encountering and how these could be solved.

5.3.4 Classroom Conversations to Support Student Inquiry

As evident in the above description, James engaged in purposeful, intentional conversations with students throughout the seven full school days committed to the hydraulics programme. Sometimes this was with the whole class, although more typically with individuals or small groups as befitting of the inquiry approach to the unit of work, both during the science investigations and then during the technological construction phase.

James' approach to teaching is based on his understanding of the importance of conversation in scaffolding students' learning, particularly when it is inquiry-based. For example, he said in an interview before the unit started:

> When it's an inquiry-based thing, the questioning is really, really key to getting them to get those good ideas – trying not to put ideas in their heads or words in their mouth, but trying to get them to come out with it, and refining it a bit.

He elaborated on this belief by identifying that the questioning

> … has to be really, really careful. You've got to step back, look at what you're doing, look at what they're doing. There've been times when I've walked over and thought, 'Oh, go away, go away! Go and think about something and come back.'

In other words, James was deliberate and thoughtful about his questioning, and very reflective about next steps, for both him as the teacher, and for the students (see Corrigan, Panizzon, & Smith, Chap. 6, this volume, for Ginny's comments about knowing when to intervene and for Heather's comments about interactive dialogue).

James' expertise as a questioner was evident in each of the lessons. Aspects that appeared critical to this expertise included:

- his commitment to valuing the students' ideas
- his belief that the students could think critically and creatively to resolve their own design problems
- his own extensive practical construction knowledge
- his own reading about hydraulic systems
- his rapport with the students, and a classroom environment of high trust.

James was also able to move fluently between scientific conceptual and procedural knowledge, mathematical conceptual and procedural knowledge, and technological conceptual and procedural knowledge, including the way hydraulics machines work.

In addition, James actively encouraged students to talk to each in order to share ideas, solve problems together, and learn from and with each other. For example, if one group was having the same problem as another group, he would encourage the two groups to talk to each other. Even the one student who worked on his own to construct his maze used the input of classmates (who were keen to have a go playing with his maze) to refine both his prototype and his final construction.

5.3.5 The Students' Learning

Figure 5.2 presents a typical example of a student group's static display showcasing their scientific investigation – the testing of the effect of syringe size on the force needed to work the hydraulic system. From the seven video explanations that were recorded by the end of day 2, six groups had clearly isolated a single variable for

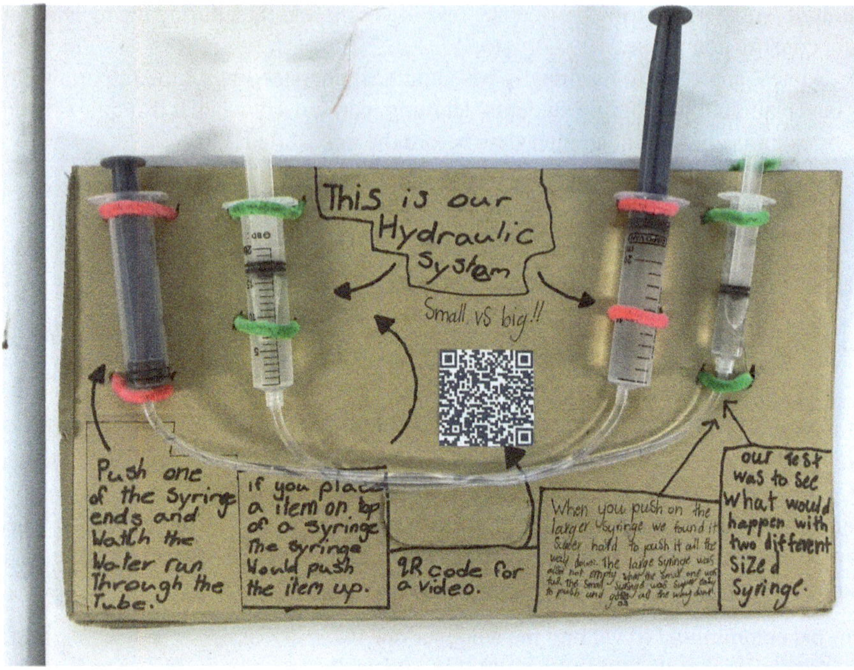

Fig. 5.2 One group's static display showing the effect of syringe size on the force required to operate the hydraulic system

testing and five groups were able to clearly articulate the results of their test. The student who worked on his own created a very engaging video that showed how a hydraulic system works, but did not show any kind of fair test being undertaken. In another case, a pair of girls working together displayed how they were testing different liquids (water and oil) but simply showed that both could be used; they didn't explore whether there were differences between the two liquids in terms of the force needed to operate the system. While these two videos were perhaps surprising, given the effort that James had given to understanding what each group was working on, they also show the value of the video reflections in understanding students' thinking in order to help them identify next steps.

In relation to the machines constructed during the second part of the hydraulics unit, Fig. 5.3 shows examples of planning documents produced by the group working on the toy retrieval arcade game. Figure 5.4 shows the prototype and final product of one of the digger projects; the second iteration was larger, and made from MDF rather than cardboard.

In the instructions for creating their final reflection videos, James specifically asked the students to critically evaluate their products and the ways in which they had worked. To ensure this was occurring, James reviewed each video before 'signing off' on it. Across the six video reflections analysed as part of this research, there was evidence of students explaining:

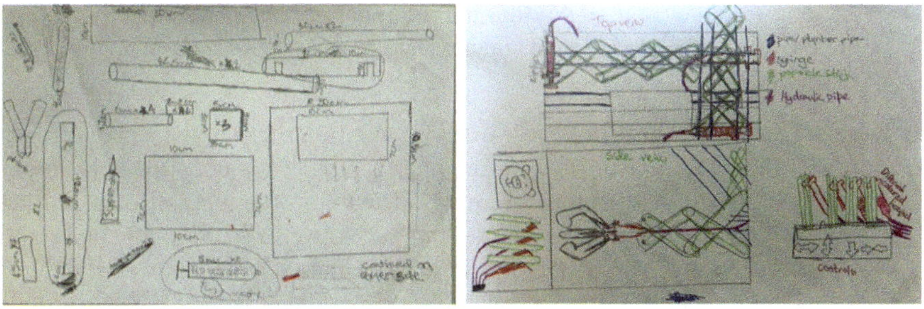

Fig. 5.3 Plans for the toy retrieval arcade game (three scissor lifts operating in two different dimensions)

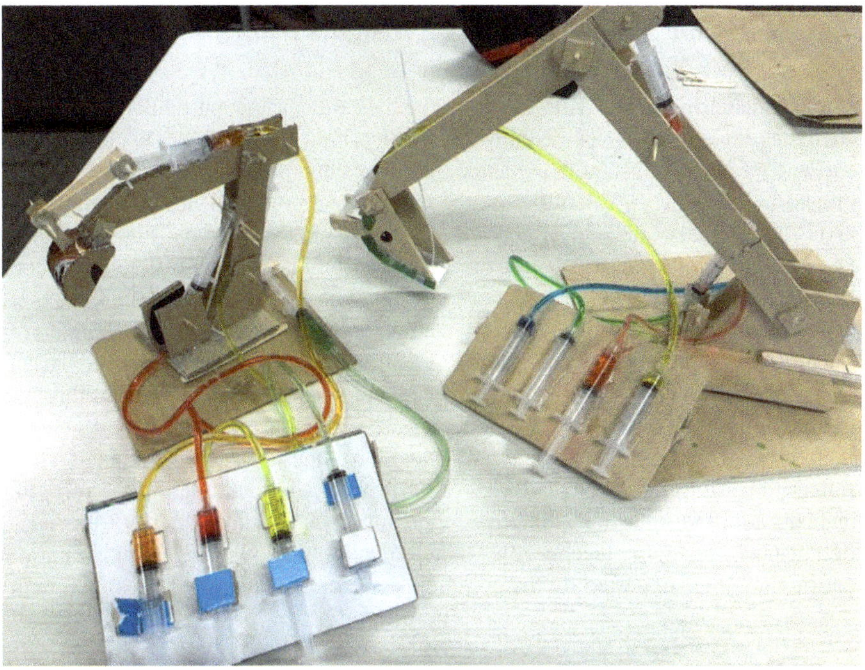

Fig. 5.4 One group's prototype (left, cardboard) and final product (right, MDF)

- what they liked about their constructions;
- challenges they had encountered and how they had resolved these;
- areas that they could have improved (e.g., working more neatly; ensuring all air had been removed from the hydraulics systems "because it would work better");
- why they had introduced changes between their prototypes and final constructions (e.g., one group introduced a swivel plate into their final construction in order for their digger arm to be able to move in different directions);

- what they would change in a future iteration (e.g., one group would "put a little more flex in the bucket because when we are trying to lift stuff up, it sometimes flies out").

Although a pair of girls indicated that they might have tried a different project altogether (their final scissor lift had too much tilt), they also pointed out ways in which they could have improved on their current construction. All other groups appeared very proud of what they had achieved, with one student concluding his group video: "This was very fun to do and I would definitely recommend it for next year" (that is, he thought that the hydraulics option should be offered to future students).

5.4 Roles for STEM Teachers – Insights from James' Class

When considering James' role(s) within the hydraulics class, particularly in light of attributes for effectively embedding science learning in rich contexts (see Fig. 5.1 earlier), we found substantial evidence that James had both the fluency and disposition to move between scientific, mathematical, technological, and everyday contexts. He was also able to effectively discern which conceptual and procedural knowledge to draw on across these domains at different points in different students' learning trajectories. These attributes came together in what we now see to be more appropriately described as *mastery* rather than competence (see Fig. 5.5). Specifically, James expertly navigated scientific, mathematical, technological and everyday discourses to support students' conceptual and skill development throughout the unit of work.

In order to further explore the ways in which James' mastery was manifest, we found it useful to refer to Crawford's (2000) case study investigating the roles that teachers adopt when using an inquiry-based approach to school science – a "myriad of constantly changing teacher roles that demand more active and complex participation than that suggested by the commonly used metaphor, teacher as facilitator"

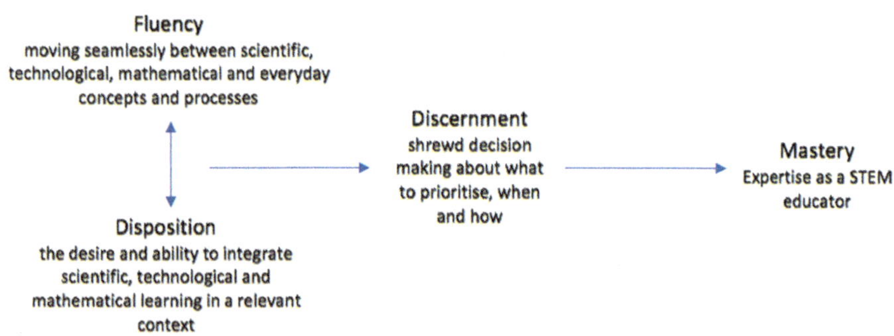

Fig. 5.5 Teacher attributes for implementing effective STEM education

(p. 935). Here, too, James was more than facilitator. While James' context was STEM inquiry rather than science inquiry, we see important consistencies with Crawford's case study. Table 5.1 lists the roles identified by Crawford, and briefly considers how these were evident in the hydraulics classroom we studied, and particularly in James' conversations.

Table 5.1 The multiple roles of a STEM educator

	Crawford's (2000) description (pp. 931–932)	Evidence in the hydraulics case study
Motivator	Encouraging students to take responsibility for their own learning	Creating an autonomous learning environment was central to James' pedagogy, evident in the inquiry-based approach to both the scientific testing and the construction phase of the unit. Through conversation, James encouraged the students to think critically and creatively when encountering problems or unexpected outcomes.
Diagnostician	Giving students opportunity to express ideas in order to discern their understandings	James' commitment to listening to students in order to probe their thinking was evident throughout the unit. Often, his responses drew on his extensive practical construction knowledge as well as the reading he had done about how hydraulic systems work – although he deliberately supported students to identify their own next steps.
Guide	Directing students and helping them develop strategies	James' responses prioritised and celebrated creativity and critical thinking when students came to him with a problem, and also when guiding them to recognise problems in their science experiments or in their technological designs and constructions.
Innovator	Designing instruction by using new ideas	James saw the potential for a hydraulics unit within the IGNITE programme goals. He didn't know how it would go, but he wanted to try it. He was very transparent with students that they were "all learning together".
Experimenter	Trying out new ways to teach and assess students	James drew on his extensive range of knowledge and skills throughout the unit. While it wasn't clear whether he was trying 'new' ways of teaching and assessing, the context (hydraulics) was certainly new.
Researcher	Evaluating one's own teaching and engaging in solving problems	Given the student-driven inquiry focus for the unit James' initial planning was very general. However, he purposefully reflected each week on the classroom learning, planning for next steps in terms of learning conversations that would be needed, as well as identifying the materials and equipment that were likely to be useful.

(continued)

Table 5.1 (continued)

	Crawford's (2000) description (pp. 931–932)	Evidence in the hydraulics case study
Modeler	Showing the attitudes and attributes [of STEM professionals] by example	James identified in an interview before the unit that he is curious by nature, and that he wants students to be curious: "For me as a person, I quite like finding out stuff, and wanting to know why. I've tried to get this through to the kids – my whole existence is inquiry based. If you want to build something, or make something, or cook something, and you don't know where to start, you've got to inquire into it – you've got to look into it, spend a lot of time researching it." In class, James regularly emphasised the importance of being able to think critically about problems, and solve them creatively.
Mentor	Supporting students in learning about [STEM] work	It was very clear that James had a strong and supportive relationship with all students in the class, and that they trusted him and felt safe about approaching him with their learning conundrums. James explicitly addressed notions of working in science, technology, architecture, and construction; and the different expertise needed within teams aiming to achieve a specific purpose.
Collaborator	Exchanging ideas with students, and allowing students to take on the role of teacher	James was committed to understanding situations from the students' point of view. He also actively encouraged students to ask each other, for example, including directing students to specific others who he had seen solving similar problems. There were many examples of ideas that arose from the students' experiences and their interpretation of these (e.g., the importance of using clean syringes for the science investigation).
Learner	Opening oneself to learning new concepts.	James clearly enjoys learning himself. He spends a considerable amount of his non-school time perusing educational sites online, pursuing hobbies, and engaging in formal professional learning opportunities. In this unit he upskilled himself in the area of hydraulics both before and during the unit. He peppered classroom conversation with examples of new things he was learning, seamlessly modelling lifelong learning.

Adapted from Crawford's (2000) description of teachers' expertise in inquiry-based science education

5.5 Concluding Thoughts

Being privy to the thinking and actions of expert teachers gives valuable insights into effective classroom practice and can help inform programmes for teacher professional learning and development. As STEM conversations become even more common-place, and some educational jurisdictions moving towards introducing formal STEM curricula, there is a need for greater understanding about how teachers might be supported to effectively plan and implement STEM programmes that meaningfully address both socio-political agendas and individual aims for education.

This representation of James' practice offers salient insights into some of the knowledge and skills needed to scaffold students' STEM learning. In particular, we draw attention to James' disposition and ability to fluently navigate multiple relevant discourses, effectively discerning which discourses were needed when and where. There is a plethora of websites that advertise the potential of simple hydraulic machines as a STEM context for learning. However the majority promote kit sets that may or may not be modifiable, with at best only very limited potential to foster creativity and critical thinking, and fewer still that explicitly introduce some of the relevant scientific, technological/engineering and mathematical concepts. Watching James and his students in action provided privileged insights into the wide range of learning that *is* possible, the learning conversations that support this, and the teacher knowledge and skills required in order for this to be the case.

Acknowledgements We are extremely grateful to James Wiltshire and his students for the warm welcome into their class.

References

Black, P. (2013). Pedagogy in theory and in practice: Formative and summative assessments in classrooms and systems. In D. Corrigan, R. Gunstone, & A. Jones (Eds.), *Assessment in science education* (pp. 207–229). Springer.

Black, P., & Harrison, C. (2004). *Science inside the black box.* GL Assessment.

Corrigan, D., Buntting, C., & Jones, A. (2012). *Rethinking the place of context in science education.* Paper presented at the annual conference of the Australasian Science Education Research Association. Wellington, New Zealand.

Crawford, B. A. (2000). Embracing the essence of inquiry: New roles for science teachers. *Journal of Research in Science Teaching, 37*, 916–937.

Jones, A. (2009). Exploring the tensions and synergies between Science and Technology in Science Education. In S. M. Ritchie (Ed.), *Handbook of research in Australasia (The world of science education Vol. 2)* (pp. 17–27). Sense.

Jones, A., & Buntting, C. (2015). Technology education and science education. In R. Gunstone (Ed.), *Encyclopedia of science education* (pp. 1049–1053). Springer.

Ministry of Education. (2014). *Introducing five science capabilities.* Retrieved from http://scienceonline.tki.org.nz/Science-capabilities-for-citizenship/Introducing-five-science-capabilities

Ministry of Education. (2007). *The New Zealand curriculum.* Learning Media. Retrieved from http://nzcurriculum.tki.org.nz/The-New-Zealand-Curriculum

Moreland, J., Jones, A., & Barlex, D. (2008). *Design and technology inside the black box.* GL Assessment.

Moreland, J., Cowie, B., & Jones, A. (2009). Assessment for learning in primary technology classrooms. *Design and Technology Education: An International Journal, 12*(2), 37–48.

Shulman, L. (1987). Knowledge and teaching: Foundations of the new reform. *Harvard Educational Review, 57*(1), 1–23.

Chapter 6
STEM, Creativity and Critical Thinking: How Do Teachers Address Multiple Learning Demands?

Deborah Corrigan, Debra Panizzon, and Kathy Smith

Abstract This chapter provides real examples that highlight how teachers must translate the concepts of creativity, critical thinking and the integrated nature of STEM in their practical realities. Such practical realities also require teachers to think about pedagogical approaches and their behaviours such as standing back with a clear pedagogical purpose, using questions to prompt student thinking and actively valuing student ideas become essential aspects of teaching practice to enhance student critical and creative thinking. Teachers also need opportunities to focus on their own thinking around these concepts by sharing and developing cumulative thinking around the nature of knowledge which defines disciplines and how to integrate this thinking with critical and creative thinking in STEM education. There is benefit in understanding creativity as a process of *producing new* ideas and critical thinking as *evaluating* and *making value judgements* in relation to evidence and arguments. In translating these concepts of creativity, critical thinking and STEM into practical realities, teachers need to consider the contexts in which they operate and look for opportunities and manage the risks that will arise. Such translations and considerations are not only difficult but are also often highly problematic in education traditions and structures that are already well-established.

Keywords Creativity · Critical thinking · STEM pedagogy · Teacher thinking

As detailed in many chapters in this volume (see, for example, Kelly & Ellerton, Chap. 2), creativity and critical thinking are seen as important competencies within the suite of twenty-first Century learning skills (OECD, 2005; p. 21), as are collaboration and communication. Creativity and critical thinking have become terms that are often linked together in educational contexts, even though they are very different concepts. Their linking comes from both being seen as important dispositions or capabilities for a future workforce.

D. Corrigan (✉) · D. Panizzon · K. Smith
Faculty of Education, Monash University, Clayton, VIC, Australia
e-mail: debbie.corrigan@monash.edu; debra.panizzon@monash.edu; kathy.smith@monash.edu

A. Berry et al. (eds.), *Education in the 21st Century*,
https://doi.org/10.1007/978-3-030-85300-6_6

We begin this chapter by considering a number of issues underpinning our thinking about creativity, critical thinking and STEM. Then we detail two examples of Australian teachers engaging with ways to develop creativity and critical thinking in STEM education. Initially we explore how these teachers, as a result of engaging in an extended professional learning programme, begin to consider STEM and its potential to develop creativity and critical thinking. We then consider what can be learnt from teaching STEM in a manner that has an intentional focus on developing students' critical thinking and creative thinking in learning areas more broadly. Finally, the chapter will draw from teachers' stories to identify the decisions that influenced the strategies, practices and approaches they used to develop students' creativity and critical thinking through STEM education and the improved learner engagement and achievement that teachers reported to have resulted.

6.1 Creativity

In framing this chapter, we have used Amabile's (1988) definition of creativity as "the production of novel and useful ideas by an individual or small group of individuals working together" (p.126) as a guide to deciding how creativity may exist in educational settings. Importantly, this definition highlights the active process required to be creative as it is the *production* of novel and useful ideas. Additionally, the definition also highlights that such production may be done by an individual or a small group. These actions associated with creativity are an important recognition that the context in which the creativity takes place will necessarily influence how it manifests. For example, it is quite common for artists to engage in creativity as individuals, as opposed to scientists who more commonly find themselves working within a team. While Amabile's definition above highlights small teams, there are many examples of large (particularly scientific) teams involved in creative endeavours; the essential element of the definition is "the production of novel and useful ideas".

Contrary to popular opinion, creative ideas are not the result of a 'lucky lightning strike' but more deliberate in nature. Regardless of the context in which creativity may be occurring, this "production of novel and useful ideas" requires some essential attributes. Creative ideas are closely linked with conscious effort, hard work and persistence (Sawyer, 2006), and depend upon quantity and output of ideas, as the more ideas you have, the more likely you are to have creative insight. For example, in schools, many educators see the activity of brain-storming as a mechanism for creative thinking. While this may represent a good start, there is often little opportunity involved in this activity for persistence and conscious effort, which are essential attributes for the development of creative competence and have been described in terms of having the capability of engaging in and acting upon creative ideas (see Kelley & Ellerton, Chap. 2, this volume). Creativity requires other attributes such as deep and technical expertise in one area and a very broad knowledge of many, seemingly unrelated other areas (Sternberg, 2006). Seeing such connections is important in developing the ability to combine disparate elements in new ways appropriate for the task or challenge at hand (Sternberg, 2006). Often creative people see patterns where others only see chaos (Sternberg, 2006). An important consideration for educators is that creativity will only happen if the creator is allowed to *fail many times in order to succeed once*.

The notion of failure in education has a dichotomous underpinning – you either fail or succeed – and is certainly not viewed as critical to, or a positive opportunity for, learning (see Mansfield & Gunstone, Chap. 9, this volume). Hence the experience of most learners is to see failure as a negative consequence of their education (such as I failed my test or I failed to submit my assignment) and to be avoided. Rarely are opportunities given to students to see failure as a positive educational outcome (for example 'my experiment didn't work and hence how can I change it?' or, 'my prototype does not meet the parameters set').

Another attribute required when engaging in creativity is the intrinsic motivation of the learners (Amabile, 1988). It is important that learners find a connection with and develop an interest in the learning context. For this reason, attending to specific problems, such as in problem-based learning, or drawing upon relevant content, becomes important. Such approaches are often used in educational settings (particularly tertiary settings) or play activities (in early childhood settings) to promote such learner investment and motivation. To use these approaches effectively, it is important that teachers know their learners to be able to tap in to such intrinsic motivation.

Hence being creative requires particular attributes to be utilised and developed if there is to be a production of novel and useful ideas. It is not just about thinking of new ideas, the active element of producing the new ideas is equally important. Creative thinking is not synonymous with creativity unless you envisage thinking as also having an action component.

6.2 Critical Thinking

The ability to think critically has four components: (i) evaluation of evidence, (ii) analysis and synthesis of evidence, (iii) drawing conclusions, and (iv) acknowledging alternative explanations and viewpoints (Council for Aid to Education, n.d.; see also Facione, 1990). Consider the first component: clearly what counts as evidence is important, and will be different across different disciplines. In science, for example, data is empirically derived, and through this process has been given value and thus is seen as evidence. In Fig. 6.1 below, data have been recorded for time to heat 500 ml of water to different temperatures. While the 'line of best fit' indicates that the temperature of the water has been rising steadily, the datum point at 30 s (highlighted by a ring around it) seems to indicate something else. Hence, given the clear pattern shown by all other datum points, we cannot place much value on the datum at 30 s (it may be a mistake) and so we ignore it as evidence. We have made a judgement about what counts as evidence when evaluating this data set.

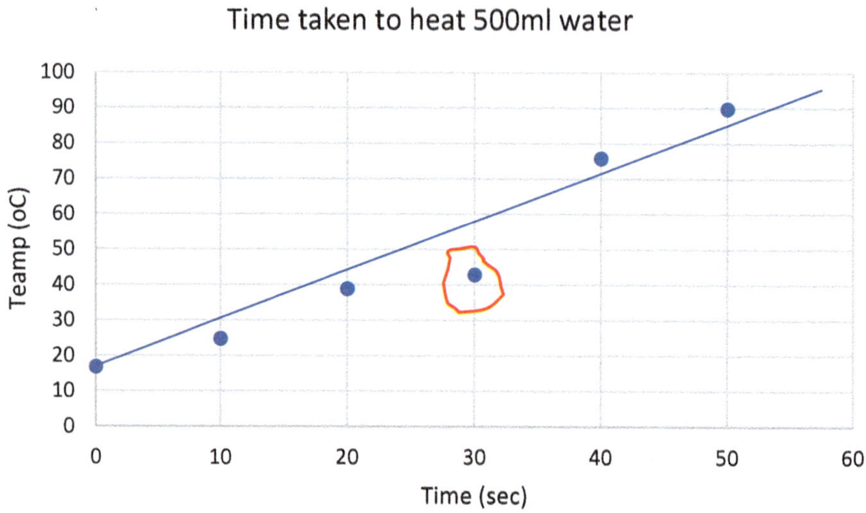

Fig. 6.1 Time taken to heat 500mls water

In comparison, data sources in history are often primary sources, such as an original document or artefact, or secondary sources, which include interpretations, analysis or commentary on the primary source. For example, a primary source could be a birth certificate, whereas a secondary source could be a commentary on the accuracy of the birth certificate. The judgements made in evaluating such a specific case rely on knowledge of process used at the time to create the birth certificate and other relevant documents (such as birth certificates of siblings or parents) to make a judgement about its accuracy.

While both these science and history examples require analysis of data and judgements to be made in evaluating data, each uses different forms of knowledge and experience in order to make such judgements. Hence the differing ways of knowing that are characteristic of different disciplines mean different (judgement) values are placed upon different data when these are evaluated. Therefore, evidence is not all the same. Additionally, while each of the two examples also requires analysis and synthesis in order to draw conclusions, the process of engagement with evidence is different in each and therefore the conclusions drawn will also be different based on different frames of thinking. Further, acknowledging different viewpoints may be appropriate; in the heating water example above an alternative point of view might consider "the line of best fit" to be inappropriate and instead conclude water indeed does not get heated at a steady rate. How to judge the quality of such alternative viewpoints is also dependent on different frames of thinking and the nature and extent of experience.

At the beginning of this section we noted that the ability to think critically has been argued to have four components. The examples above illustrate these. The four are echoed and elaborated on by Willingham (2007) when he states that critical thinking inherently involves three different types of thinking:

- Reasoning – of what types? Is it logical, deductive, inductive, rational …?
- Making judgements and decisions – based on what expertise, values, beliefs and frameworks?
- Problem solving – are the problems open or closed, convergent or divergent or networked?

While these different types of thinking will occur in all disciplines, they do look different in different disciplines, as seen from the examples above. What is also important is developing some expertise in a field and developing an awareness about what "expert" thinking looks like.

6.3 Expert Thinking and STEM

Experts have the ability to think and solve problems in ways that depend strongly on a rich body of knowledge about subject matter. They understand how facts are linked together by concepts, they recognise the underlying "big ideas" and they organise their rich knowledge base into schema. These schema allow them to recognise patterns and similarities, which in turn enables them to easily connect new information they encounter into existing schema. Expert thinking also requires metacognition; experts are aware of how they think and know what they do (Levy & Murnane, 2007).

When experts within one of the component STEM disciplines collaborate in STEM, they bring not only their expert knowledge and thinking in their own disciplines, but also a willingness to acknowledge and appreciate both their own expert thinking and that of others who hold knowledge in related and different disciplines. When educating students about STEM (STEM education), defining the nature of knowledge within each discipline becomes important. Students need to understand how the STEM disciplines are different but also related. It becomes important to understand the nature of different expertise within and across the STEM disciplines.

In tracking teachers' understanding of the different STEM disciplines, Corrigan and Smith (2020) worked with a small group (n = 20) in an online unit which was part of a post graduate course of study in STEM education. The unit was designed to support teacher leaders in STEM to explore the question 'what is STEM education?' An interactive online task was provided to explore participant teachers' thinking about the nature of the knowledge which defines each of the separate STEM disciplines. Participants were invited to share their thinking about how knowledge is defined and enacted in each discipline by adding their ideas to lists with the headings: Science, Technology/ICT,[1] Engineering and Mathematics. The contributions formed columns of ideas which remained accessible over two weeks, thus enabling

[1] ICT as defined by the Australian Office of the Chief Scientist, 2016, p.2, and so the heading "ICT" was used; see Table 6.1.

participants to read contributions and build on each other's thinking. The four columns have been brought together in Table 6.1. Each column heading denotes the discipline being discussed.

The teachers identified different ways of thinking and knowing (and acting) within these disciplines. For example, amongst participants there was a consistent view that Science and Mathematics are fundamentally concerned with explaining and representing the natural world, while Technology and ICT are more concerned with the production of an artefact using the disciplinary knowledge of science and mathematics as part of that process. Teachers also identified that disciplinary thinking in STEM requires some form of rational thinking (such as logical thinking,

Table 6.1 Teachers' views on the ways of thinking and knowing in STEM disciplines (n = 20)

Science	Technology (Engineering)	Mathematics	ICT
Takes nature apart in order to understand or explain it. Is interested in and curious about natural phenomena. Is essentially analytical in its thinking (e.g., observing data and making inferences). Is interested in being able to generalise knowledge by inventing concepts and laws and even ideal situations like ideal gases and frictionless or unbending surfaces in order to make predictions. Is often driven by a fascination or curiosity with natural phenomena. Is basically comfortable with notions like 'discovering' or 'uncovering' nature.	Puts nature together in order to make something novel. Is interested in creating and modifying artificial things. Is interested in essentially synthetic problems such as creating solutions to problems that exist. Is interested in specific knowledge; knowledge that has a bearing on a real and specific context and provides detail about a specific problem. Always has a human need or opportunity in mind. Is basically comfortable with notions of 'design' and 'invention'.	Makes sense of nature through patterns. Is interested in patterns in the natural world and representing these as models. Is logical in its thinking as it observes patterns, interprets, generalises and applies algorithms to the natural environment. Gives us models for how the world 'operates', and creates formulas/ algorithms that allow us to predict what will happen next. Is often driven by representing the natural world through algorithms and models.	Takes the mathematical patterns and applies these to make something novel or deals with large scale patterns. Uses the logic from maths extensively and creatively in generating novel applications to artificial things (e.g., digital music and art). Is interested in representing what we know about nature through coding which involves computational, algorithmic and design thinking. Programs the logic or your maths prediction. Is comfortable in taking mathematical concepts and applying these on a large scale.

analytical thinking or computational thinking), something which is often seen as characteristic of this group of disciplines.

Science and Mathematics were viewed by these teachers as having not only explanatory power but also predictive power. This view is evident in the table entries "Takes nature apart in order to understand or explain it" (for Science) and "Gives us models for how the world 'operates', and creates formulas/ algorithms that allow us to predict what will happen next" (for Mathematics). On the other hand, the power of Technology and ICT was seen to lie in the ability to solve problems and build artefacts that meet particular human needs. Table entries which illustrated such thinking include "Always has a human need or opportunity in mind" (for Technology – Engineering) and "Is interested in representing what we know about nature through coding which involves computational, algorithmic and design thinking" (for ICT).

It is also of note that many of these teachers highlighted the importance of curiosity and creativity across the disciplines. For these teachers, comparing the thinking involved between different disciplines appeared from their discussion to provide a deeper understanding of the nature of knowledge in areas beyond their specific discipline of expertise. This discussion also provided an appreciation of how working across such disciplines can open up possibilities for learning. This was an important learning experience for these teachers as few had taken the time to think deeply about the ways critical and creative thinking could be embedded in a STEM context, particularly as the Australian Curriculum[2] provides little recognition that critical and creative thinking might be expressed and conceived differently within specific subject disciplines.

6.4 Curriculum and Teachers' Practices Across Different Countries

Creative and critical thinking are not new in the curriculum although there has been a shift and renewed emphasis on these ways of thinking as part of a suite of '21st century skills' (Organisation for Economic Cooperation and Development [OECD], 2005; Silva, 2009). These two competencies are of international interest for two key reasons. Firstly, they are considered powerful in developing deeper student understanding within a discipline. Secondly, they are viewed by policymakers as ways of ensuring that citizens are able to contribute to a "modern globalised society, both economically and democratically" (Higgins, 2014, p. 566).

In many countries, creative and critical thinking are conceived as overarching goals or competencies that cut across all school curriculum areas. For example, in Northern Ireland creative thinking and personal capabilities have a central role in the Revised Northern Ireland Curriculum for students aged 4–17. The curriculum

---

[2] All teacher participants were Australian.

framework not only identifies how these capabilities might be developed over the years of schooling but also explores their relationship within other subject discipline areas. Presented in this way the competencies are perceived as 'tools' that help to deepen students' understandings of the discipline. To support teachers to embrace the intent of this framework, professional development was made available to teachers (Gallagher et al., 2012).

The New Zealand curriculum also presents 'thinking' as one of five key competencies based upon the OECD's Definition and Selection of Key Competencies [DeSeCo] project (OECD, 2005). Each competency has equal value within the curriculum. However, unlike Northern Ireland, there is no framework that positions how these competencies could be integrated into the subject discipline areas of the curriculum. This leaves it very much to the discretion of each teacher. One of the challenges identified with the New Zealand curriculum is a lack of recognition that 'thinking' might be expressed differently in various subject disciplines (Gallagher et al., 2012).

Finland has seven 'transversal competencies' that link together across subject disciplines (Halinen, 2018). 'Thinking and learning to learn' is one of the competencies that align with metacognition, and creative and critical thinking. Each of the competencies has a key statement intended to guide teachers in designing lessons and preparing learning environments that support the development of these competencies, which is explicit when compared to other curricula (Australian Curriculum, Assessment and Reporting Authority [ACARA], 2018). Even though teachers "construct their own professional guidelines based on the local curriculum" (Halinen, 2018, p. 87) these supporting statements provide direction in terms of classroom practice.

These three examples illustrate the wide international interest that exists in developing students' ways of thinking. In each example, thinking is seen as a key component of the overall curriculum framework, incorporated where relevant into specific discipline areas. It is perhaps then, not surprising that the Australian Curriculum follows a similar design. We now briefly consider relevant aspects of the Australian Curriculum since all of the teachers involved in the online unit (as described previously and below) were Australian.

6.5 The Australian Curriculum

Within Australia there is a specified national curriculum for Foundation (5 years of age) to Year 12 (16–17 years of age) that covers a range of subject disciplines. The science, mathematics, and technology curricula are structured differently although loosely around strands. These strands focus on the understandings and skills to be developed and the processes and ways of thinking comprising the nature of each discipline. Additionally, there are seven 'general capabilities' and three 'cross-curriculum priorities' to be incorporated by teachers into their practice within all discipline areas.

Critical and creative thinking together form one of the general capabilities, and are conceived as important in equipping "young Australians to live and work successfully in the 21st century" (ACARA, 2017, p. 1). Both of these components of this one capability are organised into one learning continuum that specifies the skills and attributes to be developed in learners as they progress through school. Given that critical and creative thinking are considered collectively, curriculum descriptions of creative thinking and critical thinking are indistinct and arguably of limited value to teachers in developing these different forms of thinking in their classrooms (see Vincent-Lancrin, Chap. 3, this volume).

A major challenge for all the curricula from different countries described above is that while embedding key competencies across the various subject disciplines makes sense, teachers tend to perceive them as 'add-ons' to the actual curriculum. The result is that only a few teachers embed the key competencies into their regular practice, with the majority implementing them in a tokenistic manner if at all (Gallagher et al., 2012; Higgins, 2014). This outcome is perhaps not surprising as there is no recognition from teachers that critical and creative thinking might be expressed and conceived differently within specific subject disciplines (Elder & Paul, 2010).

6.5.1 Monitoring Learner Achievement

It is an expectation that all general capabilities will be assessed as part of the Australian Curriculum. To this end, the body responsible for the Australian Curriculum, ACARA, has developed a generic continuum for critical and creative thinking that specifies a sequence of expectations around student learning and the identified key elements. Applying these expectations to student learning in a particular discipline (i.e., science, mathematics or technology) is left solely to the teachers concerned. This approach seems contradictory: on one hand individualised definitions for creative and critical thinking for each discipline are provided, and on the other a single generic continuum is given.

Until recently, this lack of alignment in the curriculum expectations has not been an issue given the lack of local and national reporting in this area. However, this is about to change, with moves towards national and international assessment of these general capabilities. Already teachers in the Australian state of Victoria are required to assess and report on student progress in critical and creative thinking, with other Australian states following suit. Furthermore, the Victorian Curriculum and Assessment Authority (VCAA), the body responsible for the Victorian-specific adaptations of the Australian Curriculum, is undertaking a large trial project involving the development and piloting of contextually-based extended response items that could be used to assess student development of this capability. Internationally, creative thinking will be included as part of the Programme for International Student Assessment (PISA) from 2021. Research suggests that such a move to a high-stakes assessment will immediately send a message to teachers and educators that critical and creative thinking are valued, thereby reinforcing the need for them to be taught and practiced by students and assessed in all discipline areas (Shepard, 2013).

6.6 Critical and Creative Thinking – Supporting Teachers Changing Their Practice

Given the move towards high-stakes assessment of these capabilities in some Australian states, the Department for Education in the state of South Australia launched a project in 2017 titled, *Critical and Creative Thinking Collaborative Inquiry Project.* This project aimed to encourage and support teachers and schools to explore the impact of an intentional focus on the development of student critical and creative thinking, in a range of curriculum learning areas. The results that emerged from this project provide some key insights into how teachers think about and work to develop critical and creative thinking in their classrooms, and the particular role that context plays in determining the pedagogies used to enhance student learning.

Four school networks across South Australia took part in this project: Network 1 involved preschools only; Network 2 involved five primary schools; Network 3 involved a combination of preschools, 2 primary schools and a single secondary school; and, Network 4 involved secondary schools only This study was site based, i.e., conducted within each education/school setting. An external evaluation of the project was undertaken to provide ongoing monitoring and analyses of data. The scope of the evaluation was to determine the impact of each network's investigation of the impact of focussing on development of critical and creative thinking, and provide recommendations for state-wide improvement in the use of strategies, practises and approaches that result in improved learner engagement and achievement. The evaluation was guided by four research questions and each question was analysed by drawing on related data sets. The Research Questions and relevant data sets were:

1. What is the nature of the inquiries implemented in sites with children and students? Data sets: *Network collaborative inquiry plans and report, which provided a brief overview of all the inquiry projects being conducted in each network, and included information about the teachers, students and children involved, the learning areas, and the timeframes for the learning.*
2. What are teachers', leaders' and educators' dispositions around critical and creative thinking? Is change identifiable from involvement in the site-based inquiries? Data sets: *Online teacher/educator survey.*
3. What kinds of strategies, tasks and/or activities are teachers and educators using to engage children and students in critical and creative thinking? Data sets: *Teacher or educator reflections completed by individuals in each network based upon their own inquiry project. Network collaborative inquiry plans and reports.*

4. What is the impact of these site-based inquiries on children and students' critical and creative thinking? Is progress or development identifiable? Data sets: *Teacher or educator reflections completed by individuals in each network based upon their inquiry project. Structured conversations. Learning story plus reflections (preschool teachers/educators). Scenarios-based assessment tasks completed by primary and secondary students. These assessment tasks were provided by VCAA.*

All data were analysed using a mix of quantitative and qualitative methods. Teacher/educator surveys were analysed using descriptive statistics, with non-parametric tests used to identify significant differences between the pre- and post data. The tests were conducted only on matched cases, i.e., individual teachers who completed both pre- and post surveys, so changes in their dispositions around critical and creative thinking were measured (Panizzon et al., 2019).

In terms of qualitative data, content analysis was used on the network collaborative inquiry plans and reports, individual teacher/educator reflections, and the structured conversation and learning story plus reflections used with preschool students. Emerging themes and insights were used to triangulate the quantitative data from the surveys in addressing each of the four research questions guiding the evaluation.

Two findings are particularly interesting in terms of teacher thinking about critical and creative thinking and the conditions created to enhance student learning: (1) defining critical and creative thinking was difficult for teachers, and (2) context shaped the pedagogical approaches teachers utilised to develop these capabilities in their students. These two findings are separately considered in the next two sections of the chapter.

6.6.1 Conflating Critical and Creative Thinking

The mixed method research findings indicated that while critical and creative thinking are general capabilities of the Australian Curriculum, defining each as a particular way of thinking appeared difficult for teachers. The Australian Curriculum documents conflate the terms, identifying four interrelated elements: Inquiring, which is about identifying, exploring and organising information and ideas; Generating ideas, possibilities and actions; Reflecting on thinking and processes; and, Analysing, synthesising and evaluating reasoning and procedures. By framing the development of critical and creative thinking in this way, that is through the interrelated elements, the learning expectations associated with each type of thinking are combined. While there is an inherent and essential interdependence between criticality and creativity (Paul & Elder, 2019), the specific nature of each type of thinking more clearly defines the intent of such thinking. As Paul and Elder (2019) explain, "creativity masters a process of making or producing, criticality a process of assessing or judging." (p. 4). However, the research we are reporting here revealed that there was not a shared understanding among the teachers about the way in which these two types of thinking converge and diverge; the teachers used the terms interchangeably. However, when participating primary teachers made judgements about student learning and achievement in relation to critical and creative thinking, their comments revealed they valued and expected two types of thinking: analytical and generative thinking. Teachers expected that an intentional focus on critical and creative thinking would improve students' capacity to analyse or interrogate information, i.e., assess the worth or validity of ideas in relation to particular contexts or needs. If a student was able to do so successfully, then teachers saw this as an indication that critical and creative thinking was taking place. Teachers also expected that a focus on critical and creative thinking would promote students' ability to demonstrate thinking that was generative in purpose, i.e., a capacity to think in ways that produced new solutions and diverse ideas. "Yet how they aligned the terms critical and creative to generative and analytical was not evident in the data" (Panizzon et al., 2019, p. 89).

6.6.2 Pedagogy: Context Matters

Each of the four school networks involved in this study chose to develop different approaches to their specific investigation. Some networks chose to initially engage teachers in shared professional development experiences in an attempt to expose teachers to pedagogical approaches designed to strategically promote critical and creative thinking, for example, thinking routines. Decisions about appropriate pedagogies were also determined by the age level of students. While play, open-ended use of materials and problem solving had always been an inherent pedagogical practice for early years educators, the explicit recognition of critical and creative thinking was not. Yet, when explicit attention was paid to these as thinking capabilities within the context of early years education, the nature of the inquiries became firmly grounded in open-ended approaches that consisted of a mix of both planned and incidental inquiries. In primary settings, problem based inquiry emerged as a popular approach for developing student critical and creative thinking. Unfamiliar or complex issues created an opportunity to explore specific problems which could engage students to find multiple solutions, promoting high levels of student intellectual engagement and active behavioural involvement while also encouraging student curiosity. Thinking routines also featured prominently in primary settings, based on the perceived value and necessity of strategic scaffolding to support student thinking. Secondary school sites faced difficulties associated with incorporating these cross curriculum capabilities within particular disciplines and traditional content structures. Most secondary teachers looked for ways of implementing critical and creative thinking into their pedagogy without distracting them or their students from curriculum requirements of specific disciplines. The majority of secondary projects undertaken by networks were practically-based around an initial building an understanding of critical and creative thinking and then the incorporating of this into classroom tasks and activities. These planning requirements and the constraint of limited time presented a challenging reality, and secondary teachers recognised the need to become more focused in what they were aiming to achieve.

6.6.3 Teachers Sharing Their Knowledge About the Conditions that Enhance Critical and Creative Thinking

This project provided evidence that a large number of teachers/educators changed their practices and pedagogies when they intentionally worked to promote critical and creative thinking more explicitly with their children and students. According to the teacher surveys conducted as part of the evaluation, this explicit attention improved children's/students' engagement, dispositions and achievement around these capabilities. The evaluation report highlighted the need for participant teachers/educators to share what they had learnt about effective pedagogies and that such professional knowledge of practice should be communicated more effectively within and across sites. There was a need to convey these insights in ways that emphasise vertical progression, i.e., capturing insights gained from early years settings, through to primary and on to secondary years, to portray a more coherent sequential narrative. It was suggested by the researchers that such information, when presented in a manner to emphasise vertical progression, may be more likely to support a child's/student's transition, growth and progression within a site in terms of critical and creative thinking.

In response to these recommendations participant teachers were invited to attend a writing workshop to share their learning in the form of 'cases' (see for example Shulman, 1992). Cases allow teachers to document their experiences of practice by capturing specific events through narrative. The purpose of the narrative is to capture contextual realities, and to take the reader beyond a description of "activities that work" (Appleton, 2002). Case writing is powerful for capturing and supporting teacher learning as it foregrounds the dilemmas teachers face and the corresponding decisions they make within the contextual reality of their teaching situations. In so doing, cases can convey new ways of thinking about teaching and learning. Writing cases helps teachers to "notice" (Mason, 2002) what they and their students do in their classrooms, thus enabling teachers to consider and confront the complex nature of teaching and learning (Loughran, 2008).

To date, five cases have been developed by participant teachers. These cases provide rich insights into the critical moments when teachers/educators became aware of the personal, professional thinking they were utilising as they worked to enhance their students' critical and creative thinking. Some excerpts are included below to highlight what teachers identified that they needed to pay attention to as they worked to support student thinking.

Roxanne, an Early Years educator who worked largely with indigenous children, found it was important to begin with building student confidence and develop in children a willingness to value their own thinking and experiences. 'Mat time'[3] presented a rich opportunity to encourage students to share, listen and extend the information they noticed in their everyday world.

> At the beginning of the term, at mat time, I always ask the children what they did in the school holidays. So, the morning came and I asked the children this question and as the microphone was passed around I realised the children were copying each other, telling the same stories, giving the same responses, no one had a different opinion. I began to wonder if I had a group of parrots? I felt that I needed to press "pause" and take a step back, I was feeling frustrated. Were they scared to share their ideas? Did they feel as though their experiences weren't good enough? Was it so risky to just say what you really did?
>
> I knew I needed to work out a way to get the children to think for themselves, think more independently and stop copying each other. I knew that some of these children did not go anywhere in the holidays but stayed at home, but this didn't explain why they always all said the same things. After talking to my colleagues, reading a few articles and thinking about this all term, I came up with a possible solution to my problem for group discussions. I needed to nudge their thinking just a little and maybe simple questions were a way forward. (Roxanne Ware)

Teachers also became aware of the need to create particular conditions to enhance learning. For Ginny, another Early Years educator, knowing when to stand back and allow time for independent student exploration became a critical teaching consideration. Early Years teachers such as Ginny began to make informed pedagogical decisions about their teaching behaviours, and to consider how these behaviours enabled students to think creatively. In doing so they were building their expertise and professional knowledge about how to develop critical and creative thinking.

> So, we learnt: Don't jump in – give them time to work it out. As educators we need to know when to intervene and what questions to ask and what alternatives to suggest. To explain further, Finn, one of the children in the centre got some crates and started to stack them to see how high he could go. He asked Shae (the teacher) to do it but she said "You can do it." He then discovered he could make a stairway by stacking them in a certain way. When he got to seven high he came to a standstill – he had another crate to use but it was clear he was trying to work out how he could get it to the top of the stack. He wandered around looking for an idea and found a wooden reel in the mud area. He could use that!! This worked and his building ended up nine crates high! Finn was talking all the time he was building and through his talk he was clarifying his ideas. When it was finished he was so proud, especially when others said: "Wow look at that – that's great Finn!"
>
> Our role was to be close by for supervision and Shae's input was to boost Finn's belief in his own ability. But she also gave him a little bit of information to extend his thinking. …. We intended the experience to promote an idea, supporting Finn to follow it through with persistence and then celebrate his achievement. He had a sense of ownership – I did that! His self-confidence exploded along with his thinking skills. (Ginny McTaggert)

In her primary classroom, Heather, a primary teacher, found interactive dialogue was essential. Her challenge was how to prompt deeper thinking in her students, and

[3] Mat time refers to the time when children sit together on a mat on the floor and attend to what their teacher is saying/doing.

the use of images and text supported with thinking routines provided a key opportunity to extend student thinking beyond merely superficial observations.

> My journey with students began with asking myself some questions: How do I encourage / prompt my students to think more deeply? I wanted students not to just notice and state the obvious but look and think deeply about a range of texts and pictures. Another important question for me was: How do I and others know what students are thinking? This question made me think of the movie "Inside Out". How was I going to get into the minds of my students and get their thoughts out, just like the emotions got out in the movie?
>
> After some deliberation, I decided to trial a thinking routine "See Think Wonder ". Three simple words but a routine that ticked all the boxes. It was a simple framework that allowed students to organise and visibly show their thinking. My hope was that it would also deepen and extend the thinking of my students, moving them beyond just noticing! (Heather Brooks)

Diversity in the ways that inquiry around critical and creative thinking was incorporated into learning environments was clearly evident in the data. Key differences were recognized between preschool teachers/educators, primary school and secondary school teachers, which is not surprising given the varied contexts, expectations and level of student experience. Secondary teachers were much more driven by curriculum requirements than the other two groups of teachers/educators. However, there were also similarities that prevailed across all sites regarding particular conditions for learning being introduced and nurtured. Teachers/educators needed to be prepared to take risks by broadening pedagogies and redefining 'success'. Children and students needed to trust that they had permission to explore new ideas, and this required teachers to develop teaching practices that provided their students with safe and supportive learning environments.

6.7 What Can We Learn from These Examples About Creative and Critical Thinking in STEM?

This chapter provides real examples that highlight how teachers must translate the concepts of creativity, critical thinking and the integrated nature of STEM in their practical realities. For example, teachers in South Australia talked about generative thinking and analytical thinking. This talk appears to be representing their practical understanding of creative thinking and critical thinking. Such practical understanding also requires teachers to think about pedagogical approaches, for example, thinking routines such as "See Think Wonder" to prompt deeper student thinking. Perhaps such frames also represent a practical approach to integrating critical thinking and creative thinking, which in turn helps to promote deeper thinking for their students. Teacher behaviours such as standing back with a clear pedagogical purpose, using questions to prompt student thinking and actively valuing student ideas became essential aspects of teaching practice to enhance student critical and creative thinking.

In the example of teachers from the STEM graduate course unit discussed earlier (see Table 6.1), sharing and developing cumulative thinking around the nature of knowledge which defines disciplines helped them to develop an understanding of disciplines for use in their practice. This experience enables teachers to think more deeply about how to integrate critical and creative thinking in STEM education.

When implementing curriculum teachers need to consider the pedagogical scaffolds to be incorporated in their teaching to realise an infused curriculum where critical and creative thinking are embedded in integrated disciplines of STEM. Hence one recommendation would be to make this curriculum development process transparent, for example via use of teacher-generated cases demonstrating pedagogical thinking and scaffolds that would assist in the development of teacher practical knowledge.

There is benefit in understanding creativity as a process of *producing new* ideas and critical thinking as *evaluating* and *making value judgements* in relation to evidence and arguments. In translating these concepts of creativity, critical thinking and STEM into practical realities, teachers need to consider the contexts in which they operate and look for opportunities and manage the risks that will arise. Such translations and considerations are not only difficult but are also often highly problematic in education traditions and structures that are already well-established.

Acknowledgements The researchers acknowledge the support from the Department for Education South Australia in funding the project discussed and permitting teacher contributions. We specifically acknowledge the Case Studies written by Ginny McTaggart, Roxanne Ware and Heather Brooks who agreed to the inclusion of identified excerpts in our chapter.

References

ACARA. (2017). *General capabilities in the Australian Curriculum: Science*. Retrieved from https://www.australiancurriculum.edu.au/f-10-curriculum/general-capabilities/critical-and-creative-thinking/

ACARA. (2018). *International comparative study: The Australian Curriculum and the Finnish National Core Curriculum*. Retrieved from https://www.australiancurriculum.edu.au/media/3922/ac-fncc-international-comparative-study-final.pdf

Amabile, T. M. (1988). A model of creativity and innovation in organisations. *Research in Organisational Behaviour, 10*, 123–167.

Appleton, K. (2002). Science activities that work: Perceptions of primary school teachers. *Research in Science Education, 32*(3), 393–410.

Council for Aid to Education. (n.d.). *Collegiate learning assessment (CLA) critical thinking, analytic reasoning, problem solving, and writing skills: Definitions and scoring criteria*. Author.

Corrigan, D., & Smith, K. (2020). Complexity, intellectual challenge and ongoing support: Key learning conditions to enhance students' engagement in STEM education. In I. Parchmann, S. Simon, & J. Apotheker (Eds.), *Engaging learners with chemistry: Projects to stimulate interest and participation* (pp. 16–45). Royal Chemical Society.

Elder, L., & Paul, R. (2010). Critical thinking: Competency standards essential for the cultivation of intellectual skills, part 1. *Journal of Developmental Education, 34*(2), 38–39.

Facione, P. A. (1990). *Critical thinking: A statement of expert consensus for purposes of educational assessment and instruction. Research findings and recommendations*. Retrieved from http://www.eric.ed.gov/ERICWebPortal/detail?accno=ED315423

Gallagher, C., Hipkins, R., & Zohar, A. (2012). Positioning thinking within national curriculum and assessment systems: Perspectives from Israel, New Zealand and Northern Ireland. *Thinking Skills and Creativity, 7*(2), 134–143.
Halinen, I. (2018). The new educational curriculum in Finland. In M. Matthes, L. Pulkkinen, C. Clouder, & B. Heys (Eds.), *Improving the quality of childhood in Europe* (pp. 75–89). Alliance for Childhood European Network Foundation.
Higgins, S. (2014). Critical thinking for 21st century education: A cyber-tooth curriculum? *Prospects, 44*(4), 559–574.
Levy, F., & Murnane, R. J. (2007). How computerized work and globalization shape human skill demands. In M. M. Suarez-Orozco (Ed.), *Learning in the global era: International perspectives on globalization and education* (pp. 158–176). University of California.
Loughran, J. (2008). Cases: Building professional knowledge of practice. In J. Loughran & A. Berry (Eds.), *Looking into practice: Cases of science teaching and learning* (Vol. 3, pp. 7–10). Monash Print.
Mason, J. (2002). *Researching your own practice: The discipline of noticing*. Routledge.
Panizzon, D., Smith, K., & Carabot, K. (2019). *Critical and creative thinking collaborative inquiry project. Final report*. Monash University.
Paul, R., & Elder, L. (2019). *The nature and functions of critical & creative thinking*. Rowman & Littlefield.
Office of the Chief Scientist. (2016). *Australia's STEM workforce*. Australian Government.
Organisation for Economic Cooperation and Development [OECD]. (2005). *The definition and selection of key competencies: Executive summary*. Retrieved from https://www.oecd.org/education/skills-beyond-school/definitionandselectionofcompetenciesdeseco.htm
Sawyer, R. K. (2006). *Explaining creativity: The science of human innovation*. Oxford University.
Shepard, L. A. (2013). Validity for what purpose? *Teachers College Record, 115*(9), 1–12.
Shulman, J. H. (1992). *Case methods in teacher education*. Teachers College.
Silva, E. (2009). Measuring skills for 21st century learning. *The Phi Delta Kappan, 90*(9), 630–634.
Sternberg, R. (2006). The nature of creativity. *Creativity Research Journal, 18*(1), 87–98.
Willingham, D. T. (2007, Summer). Critical thinking: Why is it so hard to teach? *American Educator, 31*(2), 8–19.

Chapter 7
Stimulating Creativity and Critical Thinking in Integrated STEM Education: The Contribution of Out-of-School Activities

Léonie J. Rennie

Abstract This chapter describes how effective integrated curricula with an out-of-school component encourage students to develop their STEM understanding and skills in at least three ways. First, by testing the disciplinary knowledge they have learned in real-world, authentic contexts, students come to appreciate that good understanding requires balance; that disciplinary knowledge must be complemented with interdisciplinary or integrated knowledge. Second, by investigating issues outside of the classroom, students experience a sense of the "bigger picture", enabling them to see how what they have learned can contribute to STEM-related issues beyond their classroom. Third, when students work on issues that are important to the local community and face matters relating to social values and diversity, they have opportunities to develop their senses of social and ecojustice. Three research-based examples of integrated STEM learning are analysed in terms of the OECD dimensions of creativity and critical thinking – inquiring, imagining, doing, reflecting – to illustrate how guiding students to interact with local, place-based, or community issues can benefit not only their creativity and critical thinking, but enhance their skills of communication and collaboration.

Keywords STEM integrated curriculum · Creativity · Critical thinking · Out-of-school

In thinking about the three foci for this volume – STEM (science, technology, engineering, and mathematics), creativity, and critical thinking – I felt a sense of déjà vu and wondered why. Gradually I realised that my ideas about the outcomes of education, particularly STEM education, have coalesced around two principles: the purpose of school education and the critical importance of experience. The first of these

L. J. Rennie (✉)
School of Education, Curtin University, Perth, Western Australia
e-mail: L.Rennie@curtin.edu.au

A. Berry et al. (eds.), *Education in the 21st Century*,
https://doi.org/10.1007/978-3-030-85300-6_7

principles grew from my teacher training when I was required to read Ralph Tyler's seminal exploration of curriculum (Tyler, 1949). Tyler asked four questions about curriculum and instruction in school education. "What educational purposes should the school seek to attain?" "How can learning experiences be selected which are likely to be useful in attaining these objectives?" "How can learning experiences be organised for effective instruction?" and "How can the effectiveness of learning experiences be evaluated?" These questions are sequential and in planning any curriculum the first question must be the starting point. John Wallace, Grady Venville, and I explored this question in depth, asserting "that schools should seek to provide students with the knowledge that prepares them to be responsible adults and sensible citizens in a rapidly changing global environment" (Rennie et al., 2012, p. 120). We used the term "knowledge" holistically to include the accompanying skills and capabilities. From our exploration of how integrated curriculum can contribute to this kind of knowledge – the "knowledge that counts" – we proposed a "Worldly Perspective [as] the crux to understanding the two significant dimensions of curriculum – the balance between disciplinary knowledge and integrated knowledge, and the connection between local types of knowledge and global types of knowledge" (p. 120). We emphasised that the particular curriculum context would determine the point of balance and degree of connection, and argued that offering "a curriculum that achieves both balance and connection" has the best chance of providing "students with powerful knowledge to negotiate and improve the global community in which they live" (p. 120).

The second of these principles, the importance of experience, was crystallised by an unknowingly insightful observation by our daughter Susan, who, at the age of 8 years, announced one night at the dinner table: "Today in our music class, the teacher asked some of us to sigh. But I didn't know what to do." Her older sister was astonished. "How could you not know how to sigh?" she asked, "Trixie Beldon [the central character in a series of their story books] sighs all the time!" Susan replied: "But I only *read* her sighing – I don't *see* her sighing." I used this anecdote three decades ago when asked to provide a personal summary of conference papers about gender, science and technology in primary and secondary schooling (Rennie, 1990). It illustrated perfectly the importance of actual, first-hand experience in learning science compared to the vicarious, second-hand experience of reading about it. I argued then, as I believe now, that "If pupils do not experience a meaning and a context for what they have the opportunity to learn, they are unlikely to learn it" (p. 191).

These two principles, the purpose of school education and the critical importance of experience, underpinned a statement I made at the 2008 conference of the Australasian Science Education Research Association when talking about promoting scientific and technological literacy through school-community links. I proposed that, if one goal of school education is scientific and technological literacy, then school science and technology should aim to give students a repertoire of knowledge and experiences that can be retrieved from memory to aid interpretation of new situations and provide direction for making decisions about them. I further argued that a powerful way to do this was to promote links between the school and community.

What was it about STEM, creativity, and critical thinking that triggered these memories? The explanation resides in my understanding of literacy in science and technology as fundamental to an education that prepares students for life after school. In the following sections I will overview the meanings of scientific and technological literacies then consider how they contribute to literacy in STEM and to the development of skills in creativity and critical thinking. This discussion will focus on science and technology, but in my mind the intelligent use of both is invariably intertwined with mathematics and, when artefacts or processes are involved, engineering. Next, three case studies of school students' learning in projects that involved out-of-school activities are described to illustrate the connections between integrated STEM, creativity and critical thinking. The chapter concludes with a summary of how school-community links benefit students' learning of knowledge and skills in an integrated STEM context.

7.1 Defining Scientific and Technological Literacies

The general goal of literacy in science and technology is the same as the goal of any literacy, including reading, writing, or using numbers. It is "to provide people with the tools to participate intelligently and thoughtfully in the world around them" (Pearson & Young, 2002, p. 3). For literacy in science and technology, some explanation is required about the tools people might use, and how intelligent and thoughtful participation might occur. Table 7.1 employs a framework suggested by Pearson and Young (2002) in the context of technological literacy to set out descriptions of

Table 7.1 Descriptions of Scientific and Technological Literacies (based on Rennie, 2003)

Dimension	Scientifically literate persons	Technologically literate persons
Knowledge	Are interested in and understand the world around them	Understand the designed world, its artefacts, systems, and the infrastructure to maintain them
Capability	Engage in the discourses of and about science	Have practical skills in using artefacts and fixing simple technical problems
	Are able to identify questions, investigate and draw evidence-based conclusions	Identify practical problems, design and test solutions and evaluate results
Ways of thinking and acting	Are sceptical and questioning of claims made by others about scientific matters,	Recognise risks, weigh costs and benefits associated with new technologies
		Evaluate, select and safely use products appropriate to their needs
	Make informed decisions about the environment and their own health and well-being	Contribute to decision-making about the development and use of technology in environmental and social contexts

scientific and technological literacies in three dimensions: knowledge, capability, and ways of thinking and acting. The descriptions themselves evolved in the following ways.

In a report to the Australian Government on the status and quality of school science education, Goodrum et al. (2001) argued that scientific literacy was central to quality teaching and learning. These authors offered a description of a scientifically literate person developed from a broad review of the contemporary literature, including Bybee (1997), Bybee and DeBoer (1994), Collins (1995), Fensham (1997), Jenkins (1997), Millar and Osborne (1998), the National Research Council (1996), and the OECD Program for International Student Assessment (OECD/PISA, 1999). This definition occupies the second column of Table 7.1.

Not long after the Goodrum et al. (2001) report was released, I was asked to assess the role of literacy in technology education. It seemed useful to compare and contrast it with the Goodrum et al. description of a scientifically literate person, and so describe a technologically literate person. I drew on reports from Barlex and Pitt (2000), Black and Harrison (1985, although their insightful analysis of science and technology in curriculum did not mention literacy), Jenkins (1997), Gardner, Penna, and Brass (1990), and Pearson and Young (2002) to propose the description in Column 3 of Table 7.1 (Rennie, 2003). The description of technology has close relationships with engineering (National Assessment Governing Board, 2014; Tang & Williams, 2018), and literacy in mathematics in the STEM context is often reduced to skills in numeracy (EU Skills Panorama, 2015).

The descriptions in Table 7.1 have proved useful in many contexts, most recently in an empirically-based exploration of adults' needs for literacy in science and technology (Rennie, Stocklmayer & Gilbert, 2019). In this exploration we also scrutinised the relevance of STEM, an acronym of many meanings that has often been used as a term of convenience for any one, some, or all of its component disciplines. What is its relevance here?

7.2 STEM, Creativity, and Critical Thinking

Other chapters in this volume address various aspects and meanings of STEM, but two STEM-related analyses are mentioned here to introduce STEM literacy and what are sometimes called STEM skills. This will lead us to a consideration of creativity and critical thinking.

Tang and Williams (2018) reviewed the meanings of the term "STEM literacy", together with recognised definitions of literacies in the separate STEM disciplines. They found three "similar trends and lines of inquiry in the way scientific, mathematical, technological/engineering literacies have been conceptualised" (p. 14). These were

- the creation, use and conversion of codified multimodal representations;
- the mastery of common visual resources such as annotated diagrams and geometric drawings;

- the application of cognitive and metacognitive strategies involving problem identification, planning, evaluation and self-monitoring. (p. 15)

These three trends or lines of inquiry are interrelated. While the first two have a particular focus on the communication and sharing of knowledge, all three necessitate the use of creative and critical thinking. Tang and Williams (2018) suggested that these three common trends could provide a basic – but limited – STEM literacy; "a holistic understanding of how concepts, processes and ways of thinking can be integrated and applied to the design of a solution to a real-world problem" (p. 18). However, these authors concluded that the differences among the separate disciplines in their disciplinary languages, cognitive processes, and epistemic practices limit its application, and thus a more complete conceptualisation of STEM literacy requires that these common trends be interwoven with the literacies of the separate STEM disciplines. This conclusion mirrors the argument for balance between disciplinary and interdisciplinary knowledge mentioned earlier as a means to ensure that students acquire "knowledge that counts" (Rennie et al., 2012).

We can reach a similar conclusion from another perspective by exploring what is meant by STEM skills. Siekmann and Korbel (2016) provided a review and analysis of STEM skills in the context of vocational education. Commenting on the different meanings that STEM skills would have for different groups (educators, STEM specialists, technologically proficient workers, and scientific and technologically literate citizens), Siekmann and Korbel acknowledged the difficulty of finding a common understanding. They suggested that STEM skills belong to a group of technical skills but "they overlap broadly with other skills groups such as generic and cognitive skills, as well as employability skills and the twenty-first century skills" (p. 45). They recommended against adopting the term "STEM skills"; instead it should be acknowledged "that STEM is an umbrella term" (p. 47) and that skills should be identified by their "the original definition or their category, for example, cognitive skills, foundational literacies, job-related technical skills etc." (p. 45). Siekmann and Korbel proposed that STEM skills be incorporated in a holistic skills framework, such as the Twenty-first Century Skills Framework.

The original Framework for twenty-first century Learning, developed in the United States, highlighted 18 different skills designed to be built into education standards, assessments, and professional development, but it was found to be too complex (National Education Association [NEA], n.d.). Four specific skills known as the "Four Cs" – creativity, critical thinking, communication, and collaboration – were agreed to be the most important and became the focus of education in many US states. Significantly, national educational bodies for arts, English, geography, languages, mathematics, science, and social studies, are affiliated with the Four Cs framework (NEA, n.d.). Clearly, creativity, critical thinking, communication, and collaboration are interdisciplinary skills, and this is a significant part of the argument I wish to make here.

These reviews concluded that STEM literacy and STEM skills are both umbrella terms, and both depend on the metacognitive skills of creativity and critical thinking. Here is where the sense of déjà vu kicked in. If the reader revisits Table 7.1 and peruses the descriptions of scientific and technological literate persons, it will become clear that such persons will be able to demonstrate creativity and critical thinking – no matter how these terms are defined – in the contexts of science and technology. Furthermore, the scientific and technologically literate people described in Table 7.1 will be demonstrating those skills as they participate in their day-to-day experiential world. It follows that if students are to develop these thinking skills in the context of STEM, in ways that will help them cope in later life as responsible adults and sensible citizens, they will need to practice these skills in contexts that take them outside of school.

The next sections of this chapter describe three examples of integrated STEM programmes that illustrate how students have been given opportunities to develop their creativity and critical thinking in school-community programmes. All three examples were part of larger funded projects focused on science or biotechnology but it will become obvious that all required the integration of the STEM disciplines. Creativity and critical thinking are discussed in detail elsewhere (see particularly Ellerton & Kelly, Chap. 2, this volume), but substance is given to their meaning here by using the OECD rubrics to support creativity and critical thinking in teaching and learning (Vincent-Lancrin, Chap. 3, this volume). Here, the essence of creativity is defined as "coming up with new ideas and solutions" and critical thinking as "questioning and evaluating ideas and solutions" (Vincent-Lancrin, Table 3.1). Consistent with the statements made by many curriculum authorities (see Corrigan, Pannizzon & Smith, Chap. 6, this volume), the OECD considers creativity and critical thinking to be different but closely related. For example, the Australian Curriculum describes creativity and critical thinking together with the key ideas of inquiring; generating ideas, possibilities and actions; reflecting on thinking processes; and analysing, synthesising, and evaluating reasoning procedures (Australian Curriculum Assessment and Reporting Authority [ACARA], n.d., p. 2/3). These key ideas are very similar to the OECD rubrics that provide descriptions of both creativity and critical thinking in four dimensions: imagining, enquiring, doing, and reflecting (Vincent-Lancrin, this volume). Succinct descriptions of these four dimensions can be proposed by synthesising across the OECD rubrics for creativity and critical thinking, as follows.

- Inquiring – identifying, collecting and organising information and potential approaches to problem solving;
- Imagining – generating and exploring ideas, possibilities and actions;
- Doing – creating, testing and justifying products and processes;
- Reflecting – synthesising reasoning and evaluating outcomes and procedures.

These dimensions will be used to guide analyses of three case studies. The first relates to the disposal of intractable waste and the second concerns poisonous tiger snakes[1] endemic to an urban wetland. Both were school-community projects based in primary schools. The third project involved seven students from different secondary schools who experienced a mentorship in biotechnology.

[1] Tiger snakes are a highly venomous species found in parts of Australia. They can be aggressive.

7.3 Case Study One: Disposal of Intractable Waste

The disposal of intractable waste project was part of the Science Awareness Raising Programme led by the Australian Science Teachers Association (ASTA) (Rennie & ASTA, 2003) with funding from the Australian Government's Department of Science, Education and Training. The purpose of the programme was "to promote greater understanding in the educational and broader community of why science is important, why time is spent on it in school and why scientific literacy is an important outcome of schooling" (p. viii). Seven projects across Australia were funded to field test a science awareness raising model developed by ASTA for schools to work with their community on a science-related issue of importance. The Western Australian project was based in three schools in a major rural city. The following description draws from the evaluation of the ASTA Science Awareness Raising Project (Rennie & ASTA, 2003) that reports details of the research design, data collection and analysis.

7.3.1 Context of the Project

Intractable wastes are unable to be recycled in a viable way, take considerable time to break down or are not easily destroyed, and need long-term management for community and environmental protection. At the time of the project, low-level radioactive and chemical wastes were buried in trenches at an Intractable Waste Facility at Mt. Walton, located 125 km from a major city in the goldfields of Western Australia. The facility was constructed in 1992 and is currently under care and maintenance, with no waste deposited since 2014. However, when the ASTA project began in 2002, the operating company was investigating the suitability of dumping high level radioactive waste, and this made the matter of intractable waste of considerable concern for the surrounding communities.

7.3.2 Overview of the Project Activities

Three schools undertook separate but related projects about intractable waste disposal. This case study focuses on one project involving two Year 6/7 classes (children aged 11–12 years) and their teachers. The overall aim was for students to work with community members and resources to investigate the impact of the Waste Facility on their local environment and the impact of future waste storage. The students would then inform the community about the environmental and safety considerations in moving and storing intractable waste, enabling the issues to be better understood on a broader scale.

The project was undertaken over two terms. Students began by exploring the issue of intractable waste, gathering information about the Mt. Walton site via internet and newspaper searches and interviews with stakeholders (Department of Mineral Resources, environmental and interest groups in the community) and resource people (from the museum, library, hospital, parents, and neighbours) by phone, fax, and email. They also surveyed community members about their understanding of waste disposal, particularly radio-active material; thus uncovering a range of myths and misunderstandings. Students collated and organised their findings on display boards in their classrooms. These inquiry activities served to build students' understanding of the nature of intractable waste, identify the contentious issues that underpin the problems of disposal, and try to envisage ways forward.

In the second term, students created "expert groups" to undertake more intense investigations into the critical aspects of waste disposal they had identified. Students arranged guest speakers on specific topics, visited the road leading to Mt. Walton to check its safety for transporting waste, and followed up some media reports about waste storage. They investigated the geological stability of the site of the facility and alternative uses of the waste. The costs and benefits of disposing of intractable waste were weighed up and examined together with the roles and responsibilities of federal, state, and local governments.

Progress was documented on a large display board in the main school hall to raise awareness of other students and teachers, and parents who visited the school. A Communications Officer (funded by the project) assisted students to develop the displays and provided weekly updates to inform the school community. To help raise awareness in the community, student spokespersons were interviewed on local talk-back radio and featured in stories in the newspaper. To consolidate the project outcomes, students decided to prepare an information brochure about the Intractable Waste Facility. It was checked by the Department of Environmental Protection and then distributed to the community via the school and local shopping centre.

7.3.3 Outcomes of the Intractable Waste Project

From their investigations, students developed a "big picture" perspective of the Mt. Walton Waste Facility. In terms of knowledge, students learned what intractable waste is, how it can be stored safely, and the potential environmental impact. They were exposed to different points of view from various groups, to the ethical and political responsibilities of the various levels of government and they debated relevant sensitive moral, ethical, and environmental issues. Students developed and practised skills in gathering information and assessing its credibility, organising, synthesising, and critically analysing their findings, generating ideas for the next steps and reflecting on and evaluating their progress. They learned about design relating to their poster boards, preparation and printing of the brochure, public speaking, working collaboratively, goal setting, and time management. The knowledge, skills, and understanding the children developed were merged seamlessly in an integrated STEM project in which science and technology overlapped with art, English language, health education, mathematics, and social studies.

Teachers and parents reported that motivation remained high throughout the project. Parents learned about intractable waste and the Mt. Walton Facility via their children, displays in the school and the brochure. The wider community read newspaper stories or heard the children on radio. The project evaluation revealed that students' attempts to communicate with the local community were successful; more people knew about the facility and had an increased understanding of its purpose and operation after the project than before.

7.4 Case Study Two: Living with Tiger Snakes

"Living with Tiger Snakes" was one of 24 projects across Australia in the School Community and Industry partnerships in science (SCIps) programme (ASTA, 2005), managed by the Australian Science Teachers Association (ASTA) and funded by the Australian Government. The SCIps programme aimed to forge connections between schools, their communities and industry, to enable them to collaborate on local, science-related community issues, promote scientific literacy in the school and community and enrich students' science education through project-based learning based in the students' experiential world. "Living with Tiger Snakes" was a Western Australian project involving a wildlife centre as the industry partner and a nearby primary school with its community. The available funding was used to support the Wildlife Centre Manager's time and transport for the students between the school and the Centre. The outline below is drawn from the project evaluation and details of the research design can be found in Evans, Koul, and Rennie (2007), who described the project from an environmental perspective, and Koul and Evans (2012), who analysed it as a case study of curriculum integration with a community focus.

7.4.1 Context of the Project

The Wildlife Centre overlooks a large wetland area that includes a lake, and is a significant urban habitat for highly venomous tiger snakes (*Notechis scutatus occidentalis*) that are endemic to Australia. The venom contains neurotoxins, procoagulants, and myotoxins, the effects of which can lead to renal and respiratory failure. All bites must be considered urgent and potentially lethal, and antivenom therapy is usually required. The snakes feed mainly on frogs and are important predators in an increasingly fragile wetland ecosystem. In summer, the snakes frequently invade the surrounding properties and are often killed by householders, thus threatening the wetland's biodiversity and its ecological balance. A response more appropriate than indiscriminate slaughter is to arrange for the invading snake to be "relocated", something the Centre Manager is frequently asked to do. The Living with Tiger Snakes project was designed by the Centre Manager to address the community's fear and ignorance about tiger snakes by promoting understanding of snake behaviour; precautionary and safety issues, and deterrent measures that should be taken; and appreciation of the role of tiger snakes in the ecosystem and the importance of preserving wetland diversity.

7.4.2 Overview of the Project Activities

The Centre Manager liaised with the lead teacher at a small school located in the vicinity of the lake to plan an integrated environmental programme for the school's two classes of Year 4–7 children (45 children aged 9–12 years) and their teachers. About half of the 6-week project was carried out at the Wildlife Centre and half at school. The Centre Manager led activities including nature walks around the wetland, and dipping in the lake to learn about biodiversity and lake ecology. He visited the school for lessons on food chains and food webs in the lake. A central theme addressed behavioural precautions, how to deal with snakes calmly and what to do if bitten. During the introductory and concluding sessions of the programme, snake experts brought live snakes to the Centre that the children could handle, and tiger snakes that they could look at (but not handle).

With their teachers back at school, students consolidated their learning about snake biology, behaviour, and first aid procedures. They designed a survey and then interviewed 190 community members about their attitudes and knowledge about snakes, then collated, analysed, and graphed their results. To help inform the community about snake behaviour and safety, they prepared posters, wallet cards, badges, and signage for the lake perimeter. Groups of students designed and made dioramas of snake habitats and snake unfriendly gardens. Others built displays to illustrate food webs and food pyramids. Students' in-group conversations revealed how they developed and tested their ideas to find the best ways to build their models and decide what information was most important to put on their posters.

The project focused on educating the students about tiger snakes and, through them, educating the community. To achieve this aim, the project culminated in an evening event at the Wildlife Centre, attended by about 100 family and other community members. Every child was involved in a variety of activities, including acting out role plays to demonstrate first aid in case of snake bite, power point presentations of information they had learned about tiger snakes and their survey results, and displaying and explaining their dioramas, posters, and other interpretative materials, all designed to communicate what they had learned about coping safely with tiger snakes.

7.4.3 *Outcomes of the Living with Tiger Snakes Project*

Teachers described how this integrated project contributed to all of the learning areas in the mandated state curriculum: Students learned science knowledge specific to tiger snakes, the interdependence of organisms, the environment, food chains and food webs; in technology they used the internet, power point presentations, designed and made models; in mathematics they critically analysed their survey data and worked out the best way to represent the results; in English language, children created scripts for their role plays and made verbal presentations; acting in role plays, making posters, badges, dioramas, and using computer graphics all contributed to the arts; in health and physical education, students learned about first aid and the precautions to take while bush walking. In social studies, active citizenship was demonstrated through their participation in survey data collection and presentation at the community night, contributing to a snake-safe neighbourhood. In addition, children explored the curriculum core values of social, civic, and environmental responsibility.

Tiger snakes were a recognised community issue and attendees at the evening event saw and heard a great deal about tiger snakes and safety presented in a variety of ways. Children's excitement during their presentations was a feature. One child who, in the introductory session, refused to even look at the snakes brought by the visiting herpetologists, came forward in the final session and was able to stroke a snake. Informal conversations with family members emphasised their children's excitement, interest, and learning in the topic, particularly in regard to the preservation of tiger snakes.

7.5 Case Study Three: Biotechnology Ambassadors

The World Biotech Tour (WBT) was a three-year initiative designed to promote a greater understanding of biotechnology through public outreach programmes led by science centres and museums. The WBT was coordinated by the Association of Science-Technology Centers (ASTC) based in Washington, DC and supported by the international Biogen Foundation. In 2016, Scitech Discovery Centre in Western Australia hosted the WBT and brought together students, teachers, researchers, industry personnel, and the general public to participate in a series of biotechnology-related events. The WBT included a biotechnology festival, Lab-in-a-Box (a series of specially designed biotech hands-on activities), science cafés and other discussion events, school and community outreach, and a youth ambassador programme. These components aimed to bring biotechnology to the public's attention and help them to understand its importance and social relevance. This case study is focused on the youth ambassador programme and draws from an evaluation of the WBT at Scitech (Rennie & Rennie, 2017), in which data were collected via observation and field notes; interviews with ambassadors, their parents, and mentors; ambassadors' diaries; and pre- and post-surveys. The final evaluation of the WBT provides an overview of the full programme (Boyette et al., 2018).

7.5.1 Context of the Programme

The WBT Ambassador programme aimed to increase the impact and visibility of biotechnology among a group of high school students, enabling them to experience biotechnology first hand, learn about its importance in the community, and use their learning to communicate about biotechnology with other students, their family, and wider audiences. To foster global collaboration in the programme, virtual exchange meetings were held with Ambassadors in other countries. At the end of the programme, one Ambassador was chosen to represent each country at a final world gathering in Japan in 2017.

7.5.2 Overview of the Program Activities

The Scitech WBT Manager invited academically talented students from different schools to become WBT Ambassadors. The group comprised three Year 10 students, three Year 11 students and one Year 12 student. A mentor was recruited for each Ambassador, including a pharmaceutical manager, a university science communicator, and five scientists working in biotechnology. Over a period of about five months, Ambassadors worked with their mentor outside of school to develop a project on a biotechnology topic of their choice, usually related to their mentor's occupation, and prepare a poster to synthesise their findings. The Ambassadors had five virtual exchanges with Ambassadors at science centres in Canada, Italy, and Thailand.

Part of the Ambassadors' role was to publicise the aspect of biotechnology they were researching, so to develop their communication skills, the Scitech WBT Manager led afternoon/evening sessions prior to the virtual exchanges (due to international time differences, these occurred late in the evening). These assisted Ambassadors to plan their project, design their poster, and learn how to communicate with others about their topic. Posters outlining the projects were prepared for display at the Perth Science Festival (part of National Science Week) in mid-August, where the Ambassadors interacted with visitors about their poster and demonstrated the WBT Lab-in-a-Box activities. The posters were later displayed at Scitech and the Ambassadors were present for the launch of Scitech's biotech-focused festival for schools and families, during which other WBT activities were featured. Ambassadors concluded their programme by making formal presentations about their projects at a public meeting at Scitech.

7.5.3 *Outcomes of the Biotech Ambassadors Programme*

The WBT Ambassadors programme gave seven students a unique learning opportunity to interact closely with a scientist, to pursue a particular topic in depth and learn to communicate about that topic. The Ambassadors were very able students and all produced outstanding posters and gave impressive, well-structured presentations about their project. However, each completed a different project and they had different learning journeys during their mentorships. Table 7.2 provides a flavour of their experiences by briefly describing the journeys of Hayley and Kevin (pseudonyms are used for all Ambassadors).

Despite the Ambassadors' different journeys, the data indicated clear commonalities in the outcomes of the programme. The Ambassadors quickly became comfortable with each other and enjoyed the collaborative learning that occurred during

Table 7.2 Overviews of Two Ambassadors' Journeys

Information	Hayley (Year 10)	Kevin (Year 11)
Why become an Ambassador?	Wanted a challenge for herself outside of school.	He was interested and his parents were very keen that he participate.
Project topic	Drugs used for multiple sclerosis	Misconceptions about genetic modification.
Mentor and mentoring	Pharmaceutical manager; some face-to-face meetings, also email. Preferred face-to-face.	A mature PhD student; had face-to-face weekly meetings.
Approach to topic	Mentor used brochures to explain a difficult topic so Hayley could then "translate" more difficult articles. Hayley worked fairly independently, with mentor offering advice when needed. Mentor was impressed by Hayley's explanations on her poster.	Discussion with Mentor about trying to make GM relevant to the everyday world, and exploring pros and cons. Used a survey to find out about people's understanding/ misunderstanding of GM, and was surprised to find that people had firm, often opposing, views.
Particular outcomes	Enjoyed sharing her new knowledge with others. Big increase in knowledge and confidence. Doing the research helped Hayley understand the importance of communication. Found that she learned well by talking with others.	Enjoyed doing poster, but not presentation to large audience. Enjoyed talking to small groups at the Science Festival. Big increase in knowledge and confidence, although still shy in front of a large group. Became positive about importance of communicating science and technology, and more aware of the risks involved.
Final comment from diary	"I have gained so much during the program. I learnt so much about MS and gained skills to talk to in front of people as well as communicating science to the general public."	"A great learning experience as I was able to speak with someone who is extremely knowledgeable on the subject of genetic modification and he was able to help me further my understanding of the topic."

the sessions held at Scitech prior to the virtual exchanges. They enjoyed the experience of talking to other Ambassadors during these exchanges and finding out what students in other countries were doing. Two Ambassadors were delighted to have visited their mentor's laboratory and gain some experience of what "science was really like".

Ambassadors found talking about their project with members of the public very illuminating and learned some of the basics of science communication. Anna, whose project was about epigenetic inheritance, pointed out that "my use of anecdotes and analogies did help a lot with explaining my topic", and Sheela, who studied photoreceptor cone cells and vision, wrote in her diary that she learnt "a lot about how it is easier to launch into a more technical explanation once the general idea of it has interested the audience".

All Ambassadors enjoyed the leadership role they took during their participation in the Perth Science Festival and were pleased with their final presentation, despite being nervous at the start. By preparing a poster and a presentation, two very different kinds of communication skills were learned. The Ambassadors had a specialised topic they understood in some depth and their presentations revealed how they had created imaginative and novel ways to tailor that knowledge to suit the audience to whom they were speaking. The evaluation found that this gave them considerable personal confidence and a very positive attitude towards biotechnology and its importance in today's world. As Hayley noted in her diary after her day at the Perth Festival, "It was very interesting to note how many parents wanted to know how they can get their children into the program. I guess what we are doing is very impressive."

7.6 Discussion

The case studies in this chapter provided opportunities for students to develop their literacy in science and technology as described in Table 7.1 (and, for the first two case studies, this has been demonstrated elsewhere, Rennie, 2006). Each case study illustrates the commonality between STEM, creativity and critical thinking. STEM learning occurred in the context of genuine community issues that could be described as socio-scientific, a term that privileges science, but these issues required the integration of other subject areas as the need arose. It is important to note that this facilitates a balance between disciplinary and integrated knowledge to be achieved. These experiences provided opportunities to learn and practice the skills of creativity and critical thinking in a cross-curricular way and, in terms of the Four Cs (NEA, n.d.), opportunities to develop skills in communication and collaboration.

Illustrations of the OECD's four interrelated dimensions for critical and creative thinking – inquiring, imagining, doing, reflecting – can be found in all case studies. Inquiring was the first step students took in getting to grips with their issue. The students exploring the disposal of intractable waste drew on a range of resources and selectively organised their information on wall charts as they endeavoured to understand the issue. Students shared the workload collaboratively as they brainstormed ideas about what to do next and how to test their ideas, such as inspecting the transport safety features of the road. Expert groups studied particular aspects, brought the information back to the whole class, and then, having synthesised and reflected on their findings, the students decided to design, make, and distribute a brochure as an effective way to communicate their understandings to their community.

A similar process occurred in the second school-based project. Students at the Wildlife Centre began with inquiry, learning about and looking at real snakes, and then surveying community attitudes. Their findings endorsed the need for more community understanding and tolerance of tiger snakes and a variety of means for communication – posters, dioramas, short illustrative dramas – were planned, created, and presented at the final community night. The WBT Ambassadors worked independently with a mentor to choose and explore their topic, organise their information, synthesise and analyse their findings, decide what were the key issues, and evaluate the best way to communicate them in a poster and presentation to the different audiences at a festival and a formal evening event.

Creativity, particularly in generating ideas, finding ways to organise and present information, and critical thinking in terms of analysing information, reflecting on findings, assessing progress, and deciding what information should be used and how it would be presented, were evident in each case study. Coincidentally, posters were an outcome of each project, but the creation of these, and the other means of communicating their findings, was the culmination of a collaborative process. In schools, this was group work, but even though the Ambassadors' projects were independent, they collaborated not only with their mentor and the Scitech WBT Manager, but exchanged experiences and ideas with the other local Ambassadors prior to the virtual exchanges, where they could share ideas and stories about their experiences.

As might be expected, it was noticeable that the OECD's four dimensions for critical and creative thinking are not sequential. Certainly, students started with inquiry, finding information, but from then on they moved back and forth between organising and analysing information, generating ideas, testing them by pursuing new information, reflecting on what they were doing, and evaluating progress. It is significant that there was a recognisable end point in each case study because that created a tangible focus. Students had a task; a problem to solve, and they knew when it had been achieved. This is a characteristic of most successful school-community projects (Rennie, 2006).

There is a caveat to this rosy story. Implementing a school-community programme requires a great deal of time and effort by teachers to arrange excursions, coordinate incursions from community members, and negotiate the curriculum within the available time. This requires motivation and considerable skill by teachers (Rennie, 2011). For example, teachers in the Living with Tiger Snakes project found they had to suspend some planned curriculum activities in other areas to devote more time to preparing presentations and exhibits for the community night. This required readjusting their timetable and juggling curricular priorities, and to justify this it was essential that the project became an integral part of the school curriculum, and not a time-consuming addition to it. Although the Ambassadors' schools were not involved in their projects, the students had to learn to manage their time with the demands of their school work, and this occasionally required some understanding from their teachers. Leadership from the Scitech WBT Manager was an important component of their success in achieving their goal, as was the commitment and effort of their mentors, who gave their time freely to organise meetings and assist the Ambassadors during the mentoring process.

7.7 Conclusion

The case studies demonstrated that taking students outside the classroom and providing them with learning activities that require interaction with local community issues can benefit their STEM learning and development of creativity and critical thinking in at least three ways. First, students can test the disciplinary knowledge they have learned in a real-world, authentic context. Experiencing the application of disciplinary knowledge in life outside of school demonstrates that the real world is interdisciplinary and complex, with many variables intertwined. This helps students to appreciate that good understanding requires an appropriate balance between disciplinary and interdisciplinary knowledge in order to understand and solve problems in the outside world.

Second, by connecting with issues outside of the classroom, students experience a sense of the "bigger picture". This gives context to their learning and enables them to realise that what they have learned in a local context can also have meaning in a more global context beyond their classroom. Simply learning about intractable waste in their classroom would have had little meaning to the students in the first case study. Exploring the issue in the community context gave it much more meaning, enabling students to see the social, environmental, and political ramifications. Students in the second case study learned that tiger snakes are not merely creatures to be feared and possibly killed if found in their backyard; they are creatures who have an important role in the ecological well-bearing of the wetland area. This can develop the connections between knowledge of tiger snakes in the local wetland and the concepts of ecological balance in wider, more global contexts. The WBT Ambassadors were able to explore how the laboratory-generated biotechnological knowledge in their particular project has implications for the benefit and health of the wider community, but there are also risks involved.

Third, when students work on issues that are important to their local community they have opportunities to develop their senses of social and ecojustice, as they face matters relating to social and environmental values and diversity. This was clear in both school-community projects involving the disposal of intractable waste and living with tiger snakes, and also in the biotechnology projects where ethical decisions were often at stake. Students were involved with real and recognised community issues that provided opportunities for STEM learning in context. Besides gaining relevant cognitive knowledge and skills in creativity and critical thinking, they explored values relating to environmental conservation, sustainable development, ethical behaviour, and responsibility.

These case studies of students' learning from out-of-school activities have demonstrated that in making school-community links the outcome for students was powerful knowledge. Not only did students benefit from increasing their knowledge, particularly in science and technology, in an integrated context, they developed and practiced the skills of creativity and critical thinking, and began to develop a sense of ecological and social justice. As Rennie et al. (2019) concluded from their exploration of adult learning in science and technology, these are exactly the kinds of learning activities and outcomes that build confidence and create self-directed learners.

Acknowledgements I acknowledge the valuable and competent research assistance from Gina Pearse for the Disposal of Intractable Waste Project; Rosemary Evans and Rekha Koul for the Living with Tiger Snakes project; and Brenda Gibson, Mahnoosh Milanisabzevar, and Richard Rennie for the evaluation of the World Biotech Tour at Scitech. I am very grateful to you all.

Staff at the Australian Science Teachers Association, particularly Jan Althorp, Deborah Crossing, and Helen O'Sullivan, provided wonderful support during evaluations of the Science Awareness Raising Project and the School Community Industry partnerships in science (SCIps) project. Thank you for your generous cooperation.

The World Biotech Tour (WBT) was sponsored by the Biogen Foundation and the Association of Science-Technology Centers (ASTC) in Washington, DC. Thank you to Carlin Hsueh, Manager of the WBT, and Walter Staveloz, Director of International Relations, from ASTC. Special thanks are due to Tara Broadhurst, Manager of the WBT at Scitech Discovery Centre, for her enthusiastic support, and to ASTC for permission to use data from the evaluation in this chapter.

References

Australian Curriculum Assessment and Reporting Authority [ACARA]. (n.d.). *Critical and creative thinking*. Retrieved from https://www.australiancurriculum.edu.au/f-10-curriculum/general-capabilities/critical-and-creative-thinking/

Australian Science Teachers Association. (2005). *SCIps school industry partnerships in science: Final report*. Department of Education, Science and Training.

Barlex, D., & Pitt, J. (2000). *Interaction: The relationship between science and design and technology in the secondary school curriculum*. Engineering Council.

Black, P., & Harrison, G. (1985). *In place of confusion: Technology and science in the school curriculum*. Nuffield-Chelsea Curriculum Trust.

Boyette, T., Giannakopoulou, A., & Peterman, K. (2018). *World biotech tour: Findings from an international science center and museum program*. Association of Science-Technology Centers. Retrieved from http://www.worldbiotechtour.org/final-report

Bybee, R. W. (1997). *Achieving scientific literacy: From purposes to practical action*. Heinemann.

Bybee, R. W., & DeBoer, G. (1994). Research on the goals for science education. In D. L. Gabel (Ed.), *Handbook of research on teaching and learning of science* (pp. 357–387). Macmillan.

Collins, A. (1995). National science education standards in the United States: A process and a product. *Studies in Science Education, 26*, 7–37.

EU Skills Panorama. (2015, April). *STEM skills: Analytical highlight*. Prepared by ICF and Cedefop for the European Commission.

Evans, R. S., Koul, R. B., & Rennie, L. J. (2007). Raising environmental awareness through a school-community partnership. *Teaching Science, 53*(1), 30–34.
Fensham, P. J. (1997). School science and its problems with scientific literacy. In R. Levinson & J. Thomas (Eds.), *Science today: Problem or crisis?* (pp. 119–136). Routledge.
Gardner, P. L., Penna, C., & Brass, K. (1990). Technology and science: Meanings and educational implications. *Australian Science Teachers Journal, 36*(3), 22–28.
Goodrum, D., Hackling, M., & Rennie, L. (2001). *The status and quality of teaching and learning of science in Australian schools: A research report*. Department of Education, Training and Youth affairs.
Jenkins, E. W. (1997). Scientific and technological literacy for citizenship: What can we learn from research and other evidence? In S. Sjøberg & E. Kallerud (Eds.), *Science, technology and citizenship: The public understanding of science and technology in science education and research policy* (pp. 29–50). NIFU – Norsk Institutt for Studier av Forskning og utdanning.
Koul, R. B., & Evans, R. S. (2012). Focus on community: Learning about tiger snakes at Chelsea elementary school. In L. Rennie, G. Venville, & J. Wallace (Eds.), *Integrating science, technology, engineering, and mathematics: Issues, reflections and ways forward* (pp. 100–111). Routledge.
Millar, R., & Osborne, J. (1998). *Beyond 2000: Science education for the future* (The report of a seminar series funded by the Nuffield Foundation). King's College London, School of Education.
National Assessment Governing Board. (2014). *Technology and Engineering Literacy Framework for the 2014 National Assessment of Educational Progress*. Retrieved from http://permanent.access.gpo.gov/gpo44685/ed005353p.pdf?ck=51
National Education Association [NEA]. (n.d.). *Preparing 21st Century students for a global society: An educator's guide to the "Four Cs"*. Retrieved from http://www.nea.org/assets/docs/A--Guide-to-Four-Cs.pdf
National Research Council. (1996). *National science education standards*. National Academy.
OECD Programme for International Student Assessment (OECD/PISA). (1999). *Measuring student knowledge and skills: A new framework for assessment*. OECD.
Pearson, G., & Young, A. T. (2002). *Technically speaking: Why all Americans need to know more about technology*. National Academy.
Rennie, L. J. (1990). Gender and science and technology in primary and secondary schooling: A personal summary of frame 1. In I. Granstam & I. Frostfeldt (Eds.), *Report book of the European and third world gender and science and technology conference* (pp. 190–194). Jönköping, Sweden.
Rennie, L. J. (2003, January). *Reassessing the role of literacy in technology education. Keynote address to the Third International Conference on Science*. Mathematics and Technology Education.
Rennie, L. J. (2006, August). *The community's contribution to science learning: Making it count*. Plenary address to the ACER Research Conference 2006, "Boosting science learning – What will it take", Canberra. Retrieved from https://research.acer.edu.au/cgi/viewcontent.cgi?article=1006&context=research_conference_2006
Rennie, L. J. (2011). Blurring the boundary between the classroom and the community: Challenges for teachers' professional learning. In D. Corrigan, J. Dillon, & R. Gunstone (Eds.), *The professional knowledge base of science teaching* (pp. 13–29). Springer.
Rennie, L., & Australian Science Teachers Association. (2003). *The ASTA science awareness raising model: An evaluation report* (Prepared for the Department of Education, science and training). ASTA. Retrieved from https://trove.nla.gov.au/version/217747473
Rennie, L., & Rennie, R. (2017). *Evaluation of the world biotech tour at Scitech Discovery Centre, Perth Australia, 2016* (Report to the Association of Science–Technology Centers (ASTC)), Washington DC. Unpublished report.
Rennie, L., Venville, G., & Wallace, J. (2012). *Knowledge that counts in a global community: Exploring the contribution of integrated curriculum*. Routledge.

Rennie, L. J., Stocklmayer, S. M., & Gilbert, J. K. (2019). *Supporting self-directed learning in science and technology beyond the school years*. Routledge.

Siekmann, G., & Korbel, P. (2016). *Defining 'STEM' skills: Review and synthesis of the literature–support document 2*. Adelaide, South Australia. Retrieved from https://files.eric.ed.gov/fulltext/ED570655.pdf

Tang, K.-S., & Williams, P. J. (2018). STEM literacy or literacies? Examining the empirical basis of these constructs. *Review of Education, 7*(3), 675–697. https://doi.org/10.1002/rev3.3162

Tyler, R. W. (1949). *Basic principles of curriculum and instruction*. The University of Chicago.

Chapter 8
Critical Thinking Across Disciplines in University General Education: Obesity as a Socioscientific Issue

Maurice M. W. Cheng and Jessica S. C. Leung

Abstract Arguably, we are now living in a post-scarcity era. Production is geared towards human desire rather than towards fulfilling basic needs. For the first time in human history, there are more people who are overweight than underweight. Conventional school science has often portrayed obesity as a biological problem; the way to avoid obesity is to eat a healthy diet and to lead a healthy lifestyle. Implicitly, obesity is regarded as a self-inflicted problem. Such a view, however, ignores social, political, marketing, technological, cultural and economic factors that shape an environment that determines individual eating and lifestyle patterns. This chapter reports on our university general education course that aimed to develop in students a more sophisticated view of obesity as an interdisciplinary and socioscientific issue, with the particular intention of engaging critical thinking on all these factors. We start by making the case that obesity is more than just a biological problem. A critical understanding of obesity demands thinking across disciplines. Then, we expound on our course structure and pedagogy. This is followed by a report on students' learning outcomes (n = 114) in terms of the overall changes they made in their thinking about obesity. Implications for our course development and interdisciplinary learning in the form of STEM education are also discussed.

Keywords Cross-disciplinary thinking · Issue-based learning · Conceptual change · University education · Health education

M. M. W. Cheng (✉)
Te Kura Toi Tangata School of Education, University of Waikato, Hamilton, New Zealand
e-mail: maurice.cheng@waikato.ac.nz

J. S. C. Leung
Faculty of Education, The University of Hong Kong, Pokfulam, Hong Kong
e-mail: leungscj@hku.hk

A. Berry et al. (eds.), *Education in the 21st Century*,
https://doi.org/10.1007/978-3-030-85300-6_8

8.1 Introduction

Preparing citizens who are able to make informed decisions about their lives, society and the environment is arguably a goal of science education. A strategy to achieve this goal involves the use of socioscientific issues (SSI). By contrast with teaching that portrays science as a value-free pursuit of truth, the teaching of science via SSI has the following characteristics (Zeidler, 2015, p. 998):

- Controversial and ill-structured problems that require scientific evidence-based reasoning to inform decisions about such topics.
- Deliberate use of scientific topics with social ramifications that require students to engage in dialogue, discussion, debate, and argumentation.
- Tend to have implicit and explicit ethical components and require some degree of moral reasoning.

We believe that to engage students in SSI, it is important that the issues are also relevant to their interests. With this consideration in mind, we introduced the issue of 'obesity' into a general programme available to all undergraduates at one university, and did so by positioning obesity as an issue in relation to which students should inquire about its complexity. Young adults tend to pay extensive attention to their physical bodies, which may shape a part of their self-esteem. Their interest can be reflected in the popularity of competitive reality shows such as *The Biggest Loser* (in which contestants compete to lose most weight within a given period of time, with the 'biggest loser' [of weight] becoming the winner) and in the blooming of the slimming/weight loss industry around the world. Obesity is not only a personal issue for young people, but also a phenomenon at the global level. For example, 60% of adults in OECD countries are overweight, more than 40% of these overweight adults are obese (Organisation for Economic Co-operation and Development [OECD], 2019). In short, obesity is an SSI that is pertinent to students' personal interest and is relevant for them as an issue for which citizens need to develop scientific literacy in the changing world.

8.1.1 What Causes Obesity?

From the scientific perspective, obesity refers to a situation in which body fat accumulates to the extent that it exerts adverse impact(s) on the individual's health. It is the result of prolonged positive energy balance where the energy input from food intake is larger than the energy output by the body. Excess energy is stored in the form of body fat, leading to obesity in the long term. Many hold the conception that the positive energy balance is a result of a lack of willpower in controlling one's diet and lifestyle. Studies related to students' understanding of obesity have adopted this scientific perspective (Allen et al., 2019; Ozbas & Kilinc, 2015; Weissová & Prokop, 2019). Nevertheless, when asked what causes teenage pregnancy, few would be satisfied with an explanation that is limited to our knowledge of the human reproductive system. In a similar way, biology and willpower do not provide adequate explanations of how obesity occurs or what the solutions are to obesity. For example, the worldwide prevalence of obesity has nearly tripled since 1975 (World Health Organization, 2018). Lack of willpower alone cannot explain the escalated prevalence of obesity in recent decades because there is no evidence that our willpower has changed so drastically within this short timeframe. There are factors beyond biology and willpower that contribute to our explanation of the obesity epidemic.

Food choice, for example, is not only determined by our willpower, but also by food availability, convenience, social and cultural norms, health beliefs, personal preferences, social interactions and taste. Consider one common circumstance: in some underprivileged locations in many countries, fast food outlets are more easily accessible than shops or supermarkets where nutritionally high-quality food is more readily available and affordable. Although the cost of transportation involved in shopping is not an issue for all people, it is a concern for some. Similarly, the assumption that people know how to select nutritionally high-quality food may not hold for those of low socioeconomic status (or, of course, more broadly). These people may not have the adequate education to support their knowledge and practice of living a healthy lifestyle.

Food and catering industries (and their marketing) play an important role in our lives. The food industry has developed diverse strategies to increase sales. By developing a 'bliss point' using the trio of salt, sugar and fat where the saltiness, sweetness and richness are experienced to be most appealing, the processed food industry is able to make its products irresistible to consumers. Furthermore, food advertising has permeated every aspect of daily life. Food advertisements (including for fast foods, sugared drinks and snacks) targeting children often include some kind of 'health' messages (Castonguay et al., 2013). In addition to the more traditional means of information dissemination like television and printed media, the use of digital technologies, including the Internet and mobile devices, has enabled the food industry to share unprecedented volumes of information about their products in customised messages to their consumers. Even if consumers are smart enough to identify the persuasive intent underlying such information, they may not be aware that their consumption decisions are subconsciously influenced.

The relationship between the food industry and governments is also intriguing. As a result of lobbying by the food industry, dietary advice issued by governments has never been based purely on the consideration of public health, and it continues to promote outdated research (Nestle, 2018). In market-driven economies, governments may hesitate to propose policies such as restricting the advertisement of certain food products that are potentially against the value of free markets and consumers' free choice.

In short, we suggest that obesity involves a network of complicated and interrelated causes. To address the phenomenon, it is not adequate to merely consider the biology of obesity. Stigmatising the obese is also unlikely to impact on the issue (Tomiyama et al., 2018), not only because this approach has profound moral implications, but also because it creates another hurdle for the obese to overcome before seeking appropriate support. It is important to go beyond the scientific perspective and consider how our broader social, cultural and political environments shape obesity. Causes of obesity and measures to address this phenomenon are controversial, and involve ethical considerations and a degree of moral reasoning. For these reasons, 'obesity' is an exemplar socioscientific issue that provides an opportunity for students to engage in a critical scrutiny of their thinking and of the information they come across in relation to this issue.

8.2 Critical Thinking About Obesity

Critical thinking is a recurrent theme in school education and in different disciplines in tertiary education (Davies & Barnett, 2015). We take the idea of Corrigan, Panizzon and Smith (Chap. 6, this volume) that there are four integrated components of critical thinking: (1) evaluation of evidence, (2) analysis and synthesis of evidence, (3) acknowledging alternative explanations and (4) drawing conclusions. We make two remarks on the concept of evidence in relation to exercising critical thinking in understanding obesity:

(i) Evidence both exists and is interpreted in a disciplinary matrix, and what counts as evidence varies in different disciplines. Therefore, disciplinary knowledge plays an important role in analysing, synthesising and evaluating evidence, and in drawing conclusions. Critical thinking about complex phenomena such as obesity demands the informed use of knowledge from different disciplines. This is an important issue because it reminds us of the need to consider alternative forms of evidence and hence alternative explanations.
(ii) Disciplines help us to focus on what counts as evidence. In the science discipline, energy input and output is a piece of strong evidence for the cause of obesity. Beyond the science discipline, food industry marketing strategies and the low availability of high quality food in less affluent residential areas are regarded as evidence of differing contributes to obesity. If we were to fixate only on the science discipline, we would not be able to identify other factors as evidence and hence would not be able to acknowledge causes of obesity other than those from the energy balance perspective. Suggestions for addressing obesity would then focus solely on changing individuals' eating habits and levels of physical activity.

As Toomath, an endocrinologist and past president of the New Zealand Society for the Study of Diabetes, put it when she commented on the effectiveness of dieting and doing exercise, "No other therapeutic strategy employed in medicine has such poor results… Not only was the treatment… ineffective but it [induces] a sense of guilt or hopelessness [among the obese]'" (Toomath, 2016, p. 3). This reinforces our argument that we also need to examine contributors to obesity at the societal level and the ethical considerations of treatments for the obese (Zeidler et al., 2016).

Therefore, critical thinking about obesity would involve consideration of evidence and factors from both science and other disciplines such as sociological studies and ethics (components (1) and (2) of Corrigan, Panizzon and Smith's four interrelated components of critical thinking, see Chap. 6, this volume), such that multiple explanations of the issue can be conceived of (components (3) and (4)). In this connection, we suggest there are two dimensions of thinking of thinking about, particularly, SSI that are relevant to science education, namely the technocratic dimension and the emancipatory dimension (after Femandez-Balboa, 2004). It is probable that people can engage in both dimensions of thinking about obesity. However, as we discussed above, the existing studies on students' understanding of obesity and school biology have tended to focus on only at the technocratic dimension. We are arguing such a dimension is limited.

8.2.1 The Technocratic Dimension of Obesity

The technocratic dimension of critical thinking about a socioscientific issue focuses on evaluating the rigour of scientific claims in terms of the theoretical underpinnings of the issue, and the methodologies used and the validity of the conclusions drawn based on the available scientific data. Scientific phenomena are often multicausal. When speaking about obesity, there are other contributing scientific causal factors besides excessive energy input and low energy output, issues such as biological factors like gut flora and epigenetics. In this connection, critical thinking involves evaluating the various factors or sources of evidence that are in play.

Examining obesity solely through the technocratic dimension of thinking can be likened to epistemological thinking of an absolutist nature (Kuhn, 1999), in which critical thinking involves "comparing assertions to reality and determining their truth or falsehood" (p. 24). Thinking of obesity at the personal level (that is, the level of the individual) from this dimension narrows the focus of solutions down to the accurate prescription of appropriate and balanced diets, and the design of exercise plans to suit individual needs. Technocratic considerations at the societal level are limited to estimations of costs incurred by the healthcare system and by the loss of workforce numbers and hours due to issues related to obesity, as well as to estimations of savings in healthcare expenditure that can be made through reducing the number of people with obesity. At both the personal and the societal levels, thinking within the technocratic dimension strives to attain solutions that work best (i.e., the extent to which individuals lose weight) or estimations that best fit reality (i.e., in terms of expenditure and cost saving).

Generally speaking, while the technocratic dimension of thinking acknowledges biological factors that are beyond one's control, it also views the 'fight' against obesity as one in which the obese should assume responsibility for their condition and eat less, exercise more, and live a healthier lifestyle. Specifically, obesity costs society in terms of medical and health care services, and also lost work days and productivity. Therefore, according to the technocratic view, it is important to fight against obesity as an epidemic to develop a more efficient and economically viable society. To tackle the obesity, the moral responsibility then is seen to rest on the obese (their obesity is unfair to society as society has to pay the price of consequent health problems etc.). This dimension, in general, lacks moral sensitivity towards the obese in terms of morality of justice and morality of care (Sadler, 2004; Zeidler & Keefer, 2003).

In short, the technocratic dimension has components of critical thinking – it does involve evaluation of evidence and forming an explanation of obesity. But it is based mainly on the energy balance perspective, and by extension, tends to regard obesity a result of personal-level problems. It does not consider other disciplines such as sociological studies and ethics, or their evidence and alternative explanations.

8.2.2 The Emancipatory Dimension of Obesity

The emancipatory dimension of any socioscientific issue does not preclude scientific understanding. However, this dimension has less to do with the technical examination of a phenomenon and more to do with challenging the *status quo* through an ethical and political scrutiny of the issue. The emancipatory dimension of obesity focuses on broader social institutions, and examines power relationships, inequality and social justice. These foci entail the consideration of a number of institutional factors. These include educational factors, such as whether the obese are well informed as to what it is to have a healthy lifestyle. If it is found that the obese are not well informed, the question arises as to how this educational issue should be tackled. Other institutional factors include power relationships, such as whether it is just and fair to permit direct-to-child marketing, in which commercial advertisements create associations between the promotion of nutritionally poor food and feelings of joy and fun. Finally, the socioeconomic status of people is also a factor, such as whether people living in neighbourhoods of a low socioeconomic level have easy and affordable access to nutritionally high-quality food, and whether these neighbourhoods have many fast food outlets. In this way, emancipatory thinking problematises and questions the *status quo* rather than solely interpreting obese individuals as being the problem. Such a problematisation of the *status quo* challenges us to reconsider the possibilities of creating a society that values justice, equality and moral virtues. Table 8.1 provides a summary of the technocratic and emancipatory dimensions of obesity.

We believe that both the technocratic and emancipatory dimensions are essential to science education. Thinking in the technocratic dimension through examining scientific evidence provides a unique view, but a limited view in that it only benefits from scientific understanding and reasoning. Just as science alone cannot solve all the world's problems, the technocratic dimension does not encompass all potential problems in the broader socio-political context, for example problems of equality or social justice. It is only through the emancipatory dimension, in which knowledge and evidence from other disciplines such as sociological studies and ethics are

Table 8.1 Obesity from a technocratic dimension and an emancipatory dimension

	Technocratic dimension	Emancipatory dimension
Focus	The rigour of scientific claims	The complexity of broader socio-political environments
Causes of obesity	Overeating, sedentary lifestyle, gut flora, epigenetics, endocrine disorder etc.	Institutional factors (educational factor, food industry marketing, power relationship, socioeconomic status)
Consequences of obesity	Personal health risks and their economic implication to the society	Morally inappropriate treatment of the obese
Solutions to obesity	Maintaining healthy diet and lifestyle	Re-shaping the obesogenic environment
Attitudes to obesity	Taking the obese individuals as being the problem	Problematising and questioning the status quo; reconsidering possibilities to create a society that values justice, equality and ethical-moral virtues

considered, that these problems are scrutinised. In other words, it is essential that thinking within both technocratic and emancipatory dimensions is pursued together to tackle issues such as the worldwide phenomenon of obesity.

To develop their ability to critically think about obesity as a social phenomenon, students have to appreciate that both the exact causes of obesity and precise solutions to this problem are not directly knowable, and that there musts always be a degree of uncertainty about knowledge claims (after Kuhn, 1999). Thinking critically about obesity entails not only critical scrutiny of scientific evidence, but also consideration of different types of evidence and acknowledgement of alternative explanations of the issue. Such interdisciplinary thinking involves the comparison and evaluation of judgements based on both the technocratic and emancipatory dimensions of thinking (please refer to Fig. 8.1 for a representation of critical thinking within and across these two dimensions). This requires students to focus not only on one of these dimensions, but rather to take into account different types of evidence and arguments, as well as ethical and moral considerations.

8.3 Developing the Emancipatory Dimension of Critical Thinking

The teaching we conducted about obesity took place in a university's general education course. Based on students' extremely likely forms of exposure to the media and ideas and opinions learned from their peers and views in secondary school, we assumed that the students joining our course would already have well-developed views about obesity, its contributing factors and ways to address it. Given the likely sources of their views, we predicted that their thinking prior to the course would be inclined towards the technocratic dimension. As there was little discussion either in

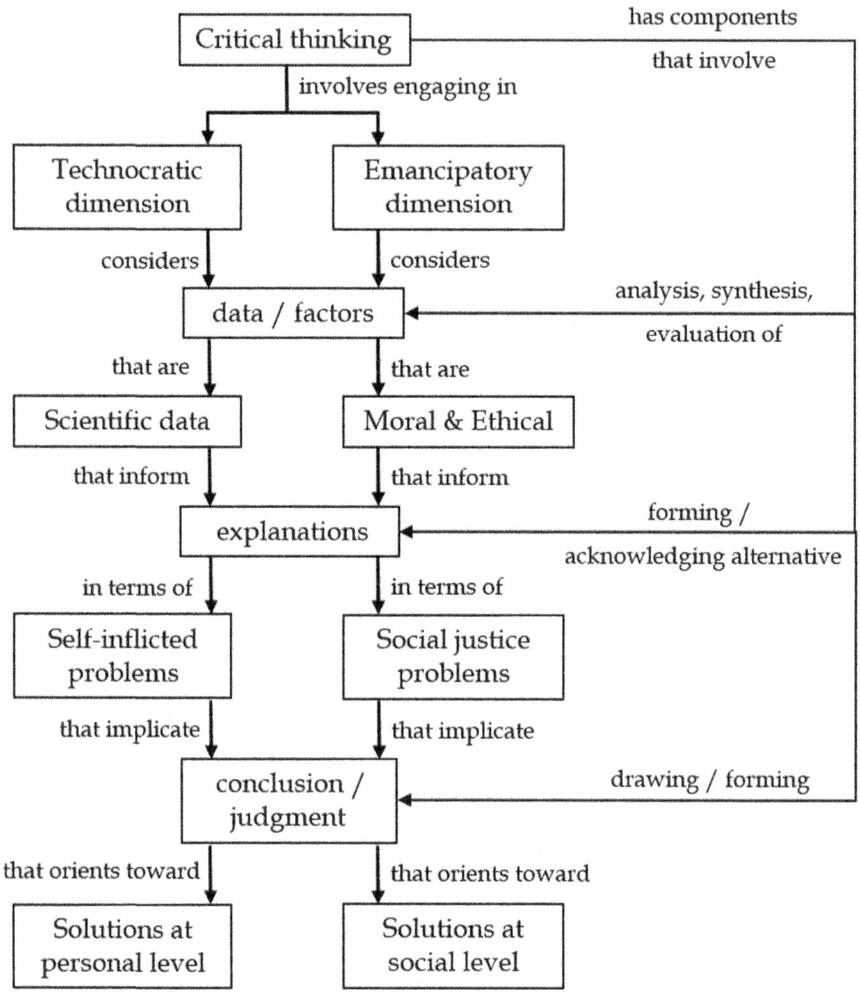

Fig. 8.1 Critical thinking involves engaging in both technocratic dimension and emancipatory dimension of the SSI, which demand thinking across disciplines

school or in the media about obesity in terms of the emancipatory dimension, our core teaching goal was to develop students' thinking in this latter dimension.

Although we sought to develop students' emancipatory thinking, it was certainly not our wish for them to completely abandon the technocratic dimension. We believed that thinking in both these two dimensions could, and should, co-exist. In what follows, we present the ideas that underpinned our belief.

1. Research studies on conceptual change and students' learning have shown that the learning of new ideas does not necessarily involve abandoning pre-existing ideas. It has been frequently shown that pre-existing and new ideas (even when

they are contradictory) can co-exist, despite students having demonstrated success in acquiring new ideas. In fact, this phenomenon occurs not only among students, but also among adults including professional scientists, as demonstrated when scientists were asked to exhibit their knowledge of a variety of science and mathematics concepts (Shtulman & Harrington, 2016).
2. Ideas that co-exist can complement each other in explaining a phenomenon. Over time, a learner might change their commitment to the pre-existing and the new ideas. Such changes in commitment depend on various factors, including the learner's recent learning experience, opportunities to make use of these ideas and different contexts where these ideas are triggered (Taber, 2019).
3. Conceptual change thus involves a shift in commitment to different ideas, rather than a replacement of one idea with another. Potvin and Cyr (2017) conceptualised these shifts in commitment as changes in different *adherence* to different ideas in specific contexts. *Adherence* to an idea is defined as the credibility status of that idea in a specific context in relation to other ideas that an individual has. In a particular context, when the *adherence* of an idea is superior to other possible competing ideas, it has a *prevalence* status. Accordingly, conceptual change is seen to involve a shift in *adherence* to various ideas and/or to involve giving *prevalence* to a particular idea in a particular moment.

Based on the above discussion of conceptual change, our teaching aimed at enhancing students' *adherence* to the emancipatory dimension of thinking, such that they would be able to develop critical thinking and evaluative judgements of issues related to obesity. In other words, we did not expect students to desist from thinking in the technocratic dimension. Rather, we were interested in shifting students' *adherence* and *prevalence* in relation to particular possible causes of obesity. In this sense, a 'conceptual change' would involve a shift from a predominantly technocratic stance about the causes of obesity to the consideration of the emancipatory dimension. This process would involve critical thinking, in which students would have to consider evidence and knowledge claims in different disciplines.

To gauge the effectiveness of our teaching, we sought answers to the following question:

> What were the changes in students' *adherence* and *prevalence* to the technocratic and emancipatory dimensions of thinking after they took our course?

We now outline our course design and then discuss how we operationalised the measurement of students' *adherence* and *prevalence*.

8.3.1 Course Design

The general education course 'Obesity: Beyond a Health Issue' was open to all undergraduate students at the university where the study was conducted. The course was an option in The General Education Programme at the university. This programme consisted of courses in four areas of inquiry (AoI): *Global issues, Scientific and technological literacy*, *Humanities* and *China: Culture, state and society*. Students were required to enrol in at least one of the courses from each AoI to fulfil their credit requirement. 'Obesity: Beyond a Health Issue' was categorised under the AoI of *Global issues* because of its emphasis on obesity as a global issue, also known as *globesity*. Most students made their course selections based on interest and schedule availability. As a general education course, our Obesity course had no science pre-requisites. This meant that students enrolled in this course might be intending majors in Arts, Business Administration, Education, Journalism, Law or the Social Sciences. This also meant that some students, those who were doing a major in Science, Medicine, Pharmacy or Engineering, would have had a background in Science, whereas others would not. Nevertheless, as our course focused on developing students' emancipatory critical thinking, we realised that a prior understanding of science, or a lack of such understanding, should not hamper their learning in this course.

The course lasted for 12 weeks. It was delivered in the form of a two-hour weekly lecture and a two-hour bi-weekly tutorial. The lectures were conducted by the second author and a professor of nutritional science from the Faculty of Science. The tutorials were conducted by lecturers from the Faculty of Science. The course design was informed by the Socioscientific Issues Teaching and Learning (SSI-TL) model of Sadler, Foulk, and Friedrichsen (2017). This model seeks to engage students in the following reasoning that is appropriate for the evaluation of both technocratic and emancipatory dimensions of thinking:

1. accounting for the inherent complexity of SSI,
2. analysing issues from multiple perspectives,
3. identifying aspects of issues that are subject to ongoing inquiry,
4. using scepticism in analyses of potentially biased information, and.
5. exploring how science can contribute to the issues and the limitations of science' (Sadler et al., 2017, p. 80).

The course structure and content are summarised in Table 8.2.

In the first unit of the course, we aimed to help students develop connections between science and the societal perspective of understanding the issue of obesity. We addressed scientific factors such as the thrifty gene hypothesis, endocrine disturbances due to sleep deprivation, epigenetics, food addiction and maternal nutrition. In units 2–6, we confronted the issue via a consideration of social, economic, cultural, political, ethical and moral factors with the intent of cultivating students' critical thinking in the emancipatory dimension. In this way, we planned to facilitate an appreciation of the complexity of the issue, in which solutions to these problems depended on *how* people framed obesity as a 'problem'. We also challenged common conceptions such as 'obese individuals are usually less healthy due to their accumulated fat' and 'significant long-term weight loss is a practical goal and will improve health', through which scepticism was exhibited in analysing potentially biased information and aspects of issues that were subject to ongoing inquiry were identified. We acknowledged that interactions among peers in different contexts

Table 8.2 The course structure of 'Obesity: Beyond a Health Issue'

Unit focuses/activity
1. Obesity: Issue overview
2. Causes of obesity: Uncovering the science of obesity (scientific perspective)
3. Causes of obesity: The plot of the multinational food industry? (marketing and political perspectives) Tutorial debate: Should soft drinks be banned at school?
4. Causes of obesity: The social construction of fat (social, cultural & economic perspectives) Tutorial debate: Should the media be responsible for fat oppression?
5. Challenging the science legitimating the battle against fatness
6. Consequences of obesity: What does fatness bring to our life and our world? Tutorial debate: Should large passengers pay for two airline tickets?
7. The way forward: Actions and attitudes towards fatness Tutorial debate: Does the fat acceptance movement encourage unhealthy lifestyles?

were key to facilitating conceptual changes and a shift in the dimension of thinking (Chi & Wylie, 2014). We therefore assigned students to engage in debates on various issues. They were encouraged to search for information on the Internet and to interpret and analyse information, and to construct arguments, counter-arguments and rebuttals based on evidence from various disciplines. These activities aimed to support students in the development of practices for making informed decisions about other SSI they may encounter in the future.

The final unit of the course was aimed at facilitating the development of students' capacity to synthesise various ideas through their engagement in a case study. Students pursued collaborative inquiries on obesity-related issues. They were free to choose issues according to their interests. The issues they chose included, but were not limited to, 'fat tax' and 'direct-to-child marketing'. We hoped that in reaching their conclusions, students would become aware of the power and limitations of science in solving these issues.

8.3.2 Measuring Shift in Dimensions of Critical Thinking

We measured students' shift in their *adherence* to and *prevalence* of the technocratic and emancipatory dimensions of thinking about obesity through the following data sources:

1. *Rating of factors contributing to obesity*

At the beginning and at the end of the course we asked students to rate their perceived importance of the contribution of different factors to obesity on a Likert scale ('5' being extremely important; '4', very important; '3', moderately important; '2', somewhat important; '1', not at all important). The factors included were regarded to be the key contributors of obesity (Foster et al., 2003; Puhl et al., 2015): (1) high fat diet, (2) overeating, (3) lack of willpower, (4) repeated dieting (weight cycling),

(5) endocrine disorder, (6) psychological problems, (7) metabolic defect, (8) genetic factors, (9) marketing/advertising of unhealthy foods, (10) poor nutritional knowledge, (11) pricing of foods, (12) physical inactivity, (13) food addiction, and (14) restaurant eating.

Factors (1) to (8) are manifestations of a technocratic dimension of thinking. More specifically, factors (1) to (4) ascribe obesity to biological factors that individuals are often thought to be able to control. Factors (5) to (8) are biological in nature but seen to be beyond the individual's control. Factors (9) to (11) address obesity at a broader societal level, and are factors that often make people of low SES more likely to gain weight. Being able to acknowledge the importance of these factors implied that the students recognised issues of social inequality within the larger issue of obesity. We thus associated these factors as manifestations of the emancipatory dimension of thinking. Factors (12) to (14) could potentially relate to either of the dimensions (e.g., low 'physical activity' may be a result either of 'laziness' or of excessive long hours of office/seat work; 'food addiction' may refer to a personal choice to indulge in food or be a result of manipulation by the food industry; 'restaurant eating' may refer to an individual's undisciplined ordering of food, or to restaurants' excessive use of fat in their dishes and their strategies of serving big portions of dishes). We thus did not categorise factors (12) to (14) as belonging to either of the two dimensions.

To determine any shift in *adherence* between the technocratic and the emancipatory dimensions, we compared the class average rating of each of these factors using a *t-test*. To determine any shift in the *prevalence* of factors that were seen to contribute to obesity, we identified the factor that received the highest class average rating in Week 1 and Week 12 of the data collection. We also identified the factors that most students decided were 'extremely important' in their rating. A comparison of these factors in Week 1 and Week 12 would reveal to us any shift in *prevalence* of factors.

2. *Guided essay writing*

This task was administered at the beginning (Week 2) and at the end of the course (Week 12). The students were required to write about causes of obesity and were asked to provide supporting arguments, counterarguments, and rebuttals (Wu & Tsai, 2007). We coded their writing based on the 14 factors of the rating task. We then compared the occasions when the students discussed these factors and used a *t-test* to determine any shift in the factors they considered. In this way, we had two data sources to determine students' shifts in *adherence* and *prevalence* about causes of obesity.

8.3.3 *Students' Learning: Shift in* Adherence *and* Prevalence *of Thinking*

Of the 120 students on the course, 116 provided consent for the use of their data; of these 114 completed the essay writing task, and 97 completed both the pre-course and post-course rating tasks.

1. *Rating task by scores*

Pre-course, and without exception, all the causes belonging to the technocratic dimension had higher scores (ranging between 3.44 for 'repeated dieting (weight cycling)' and 4.24 for 'high fat diet') than those of the emancipatory dimension (ranging between 2.73 for 'pricing of foods' and 3.36 'marketing/advertising of unhealthy foods') (see Table 8.3). This suggested a stronger *adherence* to the technocratic dimension than to the emancipatory dimension. Among all the causes, 'high fat diet' (4.24) and 'overeating' (4.19), two factors thought to be under the individual's control, were the causes of obesity that the participants *adhered* to the most. In other words, these two technocratic factors had the *prevalence* status among students at the beginning of the course.

Post-course, 'marketing/advertising of unhealthy foods' became the most *prevalent* cause (3.82, compared with 3.36 at pre-course), followed by 'high fat diet' (3.81, cf. 4.24 at pre-course) and 'overeating' (3.70, cf. 4.19 at pre-course). All the

Table 8.3 Scores of the rating task

	Pre-course		Post-course			
	M	SE	*M*	SE	*t*	*p*
Technocratic						
High fat diet	4.24	0.08	3.81	0.08	−3.61	0.00**
Overeating	4.19	0.07	3.70	0.08	−4.38	0.00**
Lack of willpower	3.48	0.09	2.99	0.10	−3.52	0.00**
Repeated dieting (weight cycling)	3.44	0.09	3.38	0.08	−0.49	0.63
Endocrine disorder	3.81	0.09	3.44	0.09	−2.88	0.00**
Psychological problems	3.75	0.08	3.47	0.08	−2.38	0.02*
Metabolic defect	3.72	0.09	3.49	0.08	−1.78	0.08
Genetic factors	3.65	0.09	3.45	0.08	−1.57	0.12
Emancipatory						
Marketing/advertising of unhealthy foods	3.36	0.12	3.82	0.09	3	0.00**
Poor nutritional knowledge	3.26	0.10	3.45	0.1	1.29	0.20
Pricing of foods	2.73	0.13	3.40	0.09	4.05	0.00**
Technocratic & emancipatory						
Physical inactivity	3.9	0.08	3.57	0.08	−2.72	0.01*
Food addiction	3.72	0.08	3.62	0.08	−0.79	0.43
Restaurant eating	3.11	0.11	3.42	0.09	2.07	0.04*

Abbreviations:
* $p < .05$
** $p < .01$

causes belonging to a technocratic dimension exhibited a decrease in *adherence*; of these the decreases of 'high fat diet', 'overeating', 'lack of willpower', 'physical inactivity' and 'psychological problems' were statistically significant ($p < 0.05$). Meanwhile, all the causes belonging to an emancipatory dimension exhibited an increase in *adherence*, with the 'increase of marketing/advertising of unhealthy foods' and 'pricing of foods' being statistically significant ($p < 0.01$). These data suggest a shift in students' thinking towards an emancipatory dimension by the end of the course. Fig. 8.2 highlights the shift in *adherence* of students' thinking from the technocratic dimension that focuses on individual responsibility to the emancipatory dimension.

2. *Guided essay writing*

Pre-course, of those causes belonging to a technocratic dimension, participants *adhered* most to 'genetic factors' (87.7%), followed by 'physical inactivity' (64.0%) and 'overeating' (62.3%) (Table 8.4). Among those causes belonging to an emancipatory dimension, the causes most *adhered* to pre-course were 'socioeconomic status' (30.7%), 'education' (18.4%) and 'marketing/advertising of unhealthy foods' (15.8%). These corroborated the result from the rating task that indicated participants strongly adhered to a technocratic view at pre-course.

Of all the causes considered by the participants at post-course, 'genetic factors' continued to be seen as the *prevalent* cause of obesity (69.3%, cf. 87.7% at pre-course), followed by 'socioeconomic status' (61.4%, cf. 30.7% at pre-course) and 'marketing/advertising of unhealthy foods' (54.4%, cf. 15.8% at pre-course). The increased *adherence* to 'socioeconomic status' and 'marketing/advertising of unhealthy foods', coupled with the *prevalent* status of 'genetic factors', suggested the emergence of an emancipatory view that coexisted with a technocratic view.

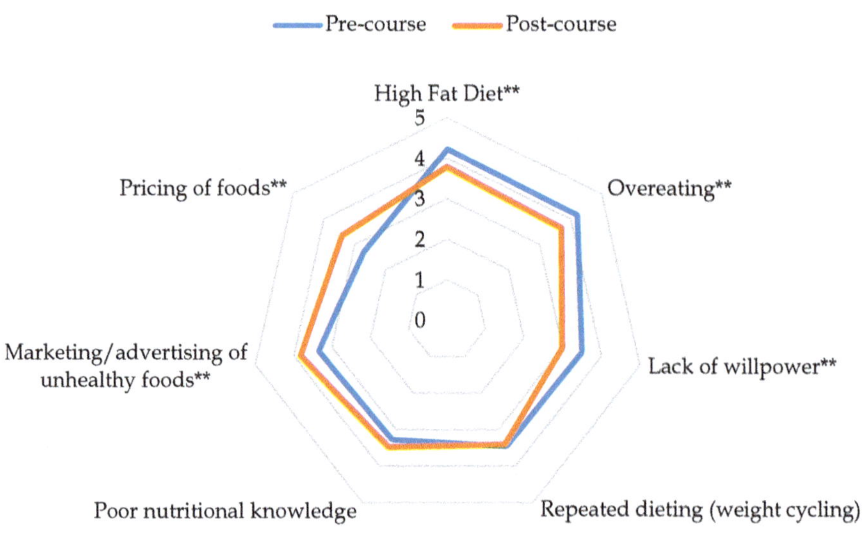

Fig. 8.2 Sources of rating tasks

Table 8.4 Causes of obesity considered by the participants in their essay writing (n = 114)

	Pre-course essay		Post-course essay		
Causes of obesity	No. of participants	% of participants	No. of participants	% of participants	% change
Technocratic					
Genetic factors	100	87.7	79	69.3	−21.0
Physical inactivity	73	64.0	53	46.5	−27.4
Overeating	71	62.3	39	34.2	−45.1
Endocrine imbalance	59	51.8	49	43.0	−17.0
Epigenetics	23	20.2	21	18.4	−8.7
Psychological problems	19	16.7	12	10.5	−36.8
Emancipatory					
Socioeconomic status	35	30.7	70	61.4	100.0
Education	21	18.4	49	43.0	133.3
Marketing/ advertising of unhealthy foods	18	15.8	62	54.4	244.4
Activity environment	9	7.9	16	14.0	77.8
Culture	8	7.0	33	29.0	312.5
Weight bias	3	2.6	30	26.3	900.0
Food lobbying	0	0.0	16	14.0	n/a
Technocratic & emancipatory					
Restaurant eating	24	21.1	10	8.8	−58.3
Food addiction	16	14.0	22	19.3	37.5

Furthermore, all the technocratic causes exhibited a decrease in *adherence* whereas all the emancipatory causes exhibited an increase in *adherence.* This further illustrated participants' shift in *adherence* from the technocratic dimension to the emancipatory dimension.

8.4 Conclusion

This chapter examines the shifts in undergraduate student *adherence* to and *prevalence* of the technocratic and emancipatory dimensions of thinking about obesity over the time of participation in a general education course on 'obesity'. In general, students exhibited a significant shift towards the emancipatory dimension. Such a shift demanded the development of critical thinking, in which students had to consider different types of evidence and alternative explanations from different disciplines. By the end of the course students demonstrated their consideration of evidence and knowledge claims beyond science, where they now also ascribed obesity to factors at the social level.

We are aware that our teaching focused on discussing causes of obesity, which left little room for students to consider measures to address obesity as a broader social phenomenon. The causes of an issue implicate the possibility of distinctive solutions, just as the means of addressing an issue are intractably linked to its causes. In our next round of teaching this course, we would like to challenge students to consider and debate measures to address obesity. We hope more students exercise emancipatory thinking with respect to the broader social, cultural and political environment and to moral reasoning about social justice and equality problems.

To end this chapter, we would like to quote a student's comment on the value of this course. It motivates us to further develop our work and to invite more students to engage in critical inquiry of SSI:

> Overall, this course provided a rewarding learning experience for me to know more about [how] individual, societal and global levels could all play a role in affecting the obesity issue. This course also enhanced my *critical thinking skills* [emphasis added] as well as the knowledge regarding obesity, it allowed me to look at the obesity epidemic in a wider lens and encouraged me to enquire more… regarding this worldwide phenomenon.

References

Allen, M., Harper, L., & Clark, Z. (2019). Preschoolers' concepts of digestive physiology and their links with body mass index. *Research in Science Education*. https://doi.org/10.1007/s11165-11019-19859-11163

Castonguay, J., McKinley, C., & Kunkel, D. (2013). Health-related messages in food advertisements targeting children. *Health Education, 113*(5), 420–432.

Chi, M. T. H., & Wylie, R. (2014). The ICAP framework: Linking cognitive engagement to active learning outcomes. *Educational Psychologist, 49*(4), 219–243.

Davies, M., & Barnett, R. (Eds.). (2015). *The Palgrave handbook of critical thinking in higher education*. Palgrave Macmillan.

Femandez-Balboa, J.-M. (2004). Emancipatory critical thinking. In J. L. Kincheloe & D. K. Weil (Eds.), *Critical thinking and learning: An encyclopedia for parents and teachers* (pp. 454–458). Greenwood.

Foster, G. D., Wadden, T. A., Makris, A. P., Davidson, D., Sanderson, R. S., Allison, D. B., & Kessler, A. (2003). Primary care physicians' attitudes about obesity and its treatment. *Obesity Research, 11*(10), 1168–1177.

Kuhn, D. (1999). A developmental model of critical thinking. *Educational Researcher, 28*(2), 16–25.

Nestle, M. (2018). *Unsavory truth: How food companies skew the science of what we eat.* Basic Books.

Organisation for Economic Co-operation and Development (OECD). (2019). *The heavy burden of obesity*. OECD Publishing.

Ozbas, S., & Kilinc, A. (2015). School students' conceptual patterns about weight gain: A preliminary study for biology teaching focusing on obesity. *Journal of Biological Education, 49*(4), 339–353.

Potvin, P., & Cyr, G. (2017). Toward a durable prevalence of scientific conceptions: Tracking the effects of two interfering misconceptions about buoyancy from preschoolers to science teachers. *Journal of Research in Science Teaching, 54*(9), 1121–1142.

Puhl, R. M., Latner, J. D., O'Brien, K., Luedicke, J., Danielsdottir, S., & Forhan, M. (2015). A multinational examination of weight bias: Predictors of anti-fat attitudes across four countries. *International Journal of Obesity, 39*(7), 1166–1173.

Sadler, T. D. (2004). Moral sensitivity and its contribution to the resolution of socio-scientific issues. *Journal of Moral Education, 33*(3), 339–358.

Sadler, T. D., Foulk, J. A., & Friedrichsen, P. J. (2017). Evolution of a model for socio-scientific issue teaching and learning. *International Journal of Education in Mathematics, Science and Technology, 5*(2), 75–87.

Shtulman, A., & Harrington, K. (2016). Tensions between science and intuition across the lifespan. *Topics in Cognitive Science, 8*(1), 118–137.

Taber, K. S. (2019). *The nature of the chemical concept: Re-constructing chemical knowledge in teaching and learning*. Royal Society of Chemistry.

Tomiyama, A. J., Carr, D., Granberg, E. M., Major, B., Robinson, E., Sutin, A. R., & Brewis, A. (2018). How and why weight stigma drives the obesity 'epidemic' and harms health. *BMC Medicine, 16*(123), 1–6.

Toomath, R. (2016). *Fat science: Why diets and exercise don't work - and what does*. Auckland University.

Weissová, M., & Prokop, P. (2019). Alternative conceptions of obesity and perceptions of obese people amongst children. *Journal of Biological Education*. https://doi.org/10.1080/00219266.00212019.01609549

World Health Organization. (2018, February 16). Obesity and overweight. *Fact sheet*. Retrieved from http://www.who.int/mediacentre/factsheets/fs311/en/

Wu, Y. T., & Tsai, C. C. (2007). High school students' informal reasoning on a socio-scientific issue: Qualitative and quantitative analyses. *International Journal of Science Education, 29*(9), 1163–1187.

Zeidler, D. (2015). Socioscientific issues. In R. F. Gunstone (Ed.), *Encyclopedia of science education* (pp. 998–1003). Springer.

Zeidler, D. L., Herman, B. C., Clough, M. P., Olson, J. K., Kahn, S., & Newton, M. (2016). Humanitas emptor: Reconsidering recent trends and policy in science teacher education. *Journal of Science Teacher Education, 27*(5), 465–476.

Zeidler, D. L., & Keefer, M. (2003). The role of moral reasoning and the status of socioscientific issues in science education: Philosophical, psychological and pedagogical considerations. In D. L. Zeidler (Ed.), *The role of moral reasoning on socioscientific issues and discourse in science education* (pp. 7–38). Kluwer Academic.

Chapter 9
When Failure Means Success: Accounts of the Role of Failure in the Development of New Knowledge in the STEM Disciplines

Jennifer Mansfield and Richard Gunstone

Abstract Often failure is not inconsistent with success, in both the advancement of the disciplines of Science, Technology, Engineering and Maths (STEM) and in the teaching and learning of these disciplines in education contexts. This chapter specifically seeks to explore the ways failure is represented in each of the STEM disciplines, and through this to infer the role and nature of failure in the development of new knowledge in each discipline. We start by discussing the notion and variety of failures, why failure is often perceived negatively, yet is an essential element of the learning process. The nature of failure in each of the STEM disciplines is explored in turn. In science, failure is commonplace as science is essentially driven by a desire to understand the world around us. Science can be context independent rather than design focused. Therefore, the end product that is communicated consists of the knowledge generated and the 'successful' process that led to it. Failures are important aspects of the process, but are seldom considered desirable or worth publishing. We contrast with the role of failure in engineering and technology, where failure is celebrated as being an integral part of the design process and demonstrates rigour of the testing and process. Failure in maths involves both certainty and failure in its quest for a solution. The fundamental premise of maths could be argued to include finding a solution to a problem or developing skill as compared to focusing heavily on generating knowledge for the sake of generating knowledge per se. The role of failure in school learning of STEM disciplines is considered briefly.

Keywords Failure · Success · STEM · Science · Design process

J. Mansfield (✉)
Faculty of Education, Monash University, Clayton, VIC, Australia
e-mail: jennifer.mansfield@monash.edu

R. Gunstone
Monash University, Clayton, VIC, Australia
e-mail: richard.gunstone@monash.edu

A. Berry et al. (eds.), *Education in the 21st Century*,
https://doi.org/10.1007/978-3-030-85300-6_9

9.1 Introduction

In the everyday world, the common view of failure is that it has no merit and does not lead to anything of worth. Yet failure is often consistent with success. This is apparent in both the advancement of the disciplines of Science, Technology, Engineering and Maths (STEM) and in the teaching and learning of these disciplines in education contexts. This chapter specifically seeks to explore the ways failure is represented in each of the STEM disciplines, and through this to infer the role and nature of failure in the development of new knowledge in each discipline.

The way failure is used and valued impacts on the ways the discipline is represented to and perceived by those outside the discipline (see, for example, Manalo & Kapur, 2018). Thus, these representations impact on the ways people then perceive and relate to failure within education contexts that involve these disciplines. This chapter considers the role of failure in each of the separate S, T/E and M disciplines. With T and E considered to be sufficiently similar in terms of the fundamental epistemologies of each, and so the nature and recognition of 'failure' in success in each, that these are considered together. The chapter concludes with a brief consideration of the role of failure in school learning of the STEM disciplines.

The purpose of this chapter is to discuss if and how failure is represented as integral to the development of new knowledge in each S, T/E and M discipline, and to then consider how the public illustrations of, and value seen for, failure are often not aligned with these representations. This purpose has implications for how failure in these disciplines is then valued and articulated in education settings. It is outside the scope of this chapter to seriously explore the ways in which failure is interpreted and portrayed in school-based versions of the STEM disciplines. Instead, we briefly consider the ways in which teachers could start to develop awareness of the value and role of failure in the separate disciplines to bring into alignment the ways in which failure is represented in the STEM education disciplines and make links with the themes of creativity and critical thinking that are central to this book (see Ellerton & Kelly, Chap. 2, this volume).

Before our considerations of the way failure is represented in each of the separate disciplines, it is important to first consider in some detail how the notion of 'failure' is more generally understood, perceived and used.

9.2 What Is Failure?

> "It is impossible to live without failing at something … unless you live so cautiously that you might as well not have lived at all – in which case you fail by default." J K Rowling

Failure, or the lack of success,[1] is often perceived as something undesirable, particularly in an era where we are constantly measured and ranked according to our

---

[1] The first synonym given for "failure" in Roget's thesaurus is "non-success"

successes and failures. In schools and universities, students are measured and ranked by the extent to which they have successfully answered questions on tests. These successes are further translated into ranking scores (e.g., Australian Tertiary Admission Ranking—ATARs, or Grade Point Averages—GPAs) which are used as commodities of academic quality and thus as leverage into further study or employment. In schools, failure to spell a word correctly eliminates you from a spelling bee competition. In the sporting field, failure to perform well in sport trials means you are less likely to be picked for a sporting team. Getting too many questions wrong on a driver's licence test means you are deemed unable to drive a car. In academia, scholars are measured and ranked on metrics (number of publications, student satisfaction rankings, number of successful grants, etc.), with higher numbers being perceived as indicators of higher proficiency. Indeed, universities jostle for positions on global ranking scales, based on the successes of their academic staff. Throughout our lives, failure and success are used as means of measuring those who can against those who can do better. It is no wonder that we are conditioned to see failure as being undesirable. Yet it is our journey of failures that can often lead to growth and improvement, thus leading to greater success.

This chapter aims to consider the representation of failure in each of the S, T/E and M disciplines and its value and importance for growth, development and advancement of that discipline. We argue that the ways in which failure is spoken about outside a discipline are not necessarily representative of the role failure plays within the discipline.

We begin by considering the meanings associated with 'success' and 'failure'. Then we elaborate some aspects of the general ways 'failure' is represented in science, technology/engineering and mathematics, and briefly consider failure as a path to learning. Later in the chapter we consider specifically how failure is perceived and represented in the formal and informal literatures about the nature and development of knowledge in each of the separate STEM disciplines, and the flow on effect of these representations into primary and secondary schooling. We argue that failure is an integral and essential aspect of knowledge development, particularly in many of the STEM disciplines, yet is rarely represented as such in some of those disciplines. This has implications for the validity of the ways in which significant aspects of the nature of some STEM disciplines are represented in schools. This can lead to a lack of appreciation of failure as a necessary ingredient in growth and development.

9.3 Defining Success and Failure

"Every failure is a stem to success." William Whewell

To succeed or be successful at something is based on the capacity to achieve a "favourable or desired outcome" (Success, n.d.). An idealised or desired state is often used as a benchmark for measuring success or degrees of success. Failure, the antithesis of success, describes moments where success is not achieved, that is, the "omission of occurrence of performance" or "falling short or being deficient" (Failure, n.d.). Portrayed in this way, it is easy to see why failure is generally perceived as negative and undesirable. Who would be satisfied with being seen as 'not performing as expected' or to 'fall short' of some idealised benchmark?

Firestein (2015) emphasises that "like so many important words, failure is much too simple for the class of things it represents. Failure comes in many flavours, and strengths, and contexts, and values, and innumerable other variables" (p. 7). This variety is represented through terms like error, mistake, blunder, faux pas, misstep, botch, disaster, let-down, catastrophe; the list goes on. These words exist to differentiate between, and categorise, different aspects of the process of striving for success. To illustrate the breadth of this diversity, Firestein (2015) suggests a *continuum of failures*, from simple lessons (e.g., 'take more care next time') to larger 'character building' failures, and from small and easily dismissed failures, to large, catastrophic and harmful failures. Firestein's list does not claim to be exhaustive, but to more simply offer an exemplar range of failure types (see Table 9.1).

The variety of failures suggested by Firestein (2015) highlights that while some failures are undesirable and unavoidable, for the most part failures can lead to learning. Smaller 'blunders' or avoidable failures can be described as mistakes or errors and can occur due to incompetence or lack of judgement which is considered 'wrong'. Examples include things like incorrect placement of a decimal point in a mathematics problem, saying the wrong thing at an inappropriate moment or dropping a coffee mug and spilling its contents on the floor. As simple or trivial as these 'errors' may sometimes seem, at times these result from some fundamental lack of knowledge as to the socially appropriate way of acting, or a muscular problem, or an inadequate conceptual understanding.

The essential point of the examples in the paragraph above is that there are many different ways of defining the nature and degree of failure (e.g., trivial to

Table 9.1 Examples of the diversity of failures

Waste of time errors or mistakes, e.g., can arise from stupidity, indifference, naivety
Errors from which we learn simple lessons, e.g., take more care next time, check the answers more carefully
Painful, character building errors which lead to life lessons, e.g., failed marriage, failed business venture
Failures that lead to unexpected discoveries, e.g., serendipity, accidental failure
Failures that are informative, e.g., does not work one way, must work another
Layers of failure: Failures which pile up and learning is related to what does not work compared to what does work
Failures that were successes for a while then were not, e.g., alchemy
Stein failure, e.g., failures which leave a wake of interesting stuff behind
Failures that are an end in themselves
Catastrophic failures

Adapted from Firestein (2015)

catastrophic), as well as the degree of seriousness of the products of failure (e.g., near misses to high numbers of casualties).

Regardless of how it is communicated, failure is integral to the process of learning. In its many forms, failure helps us recognise something new, unexpected and valuable. It can help us learn how to (or how not to) behave in the future, if only we recognised the lesson. Viewed in this way, failure is valuable and inextricably linked to the development of knowledge (in an individual or in a whole domain of knowledge), as John Dewey noted in one of his many oft-quoted epithets: "Failure is instructive. The person who really thinks learns quite as much from his failures as from his successes" (Dewey et al., 2008, p. 206). Yet despite these powerful and realistic arguments, failure is often labelled with pejorative terms and discussed using negative prose; failure is often perceived as something undesirable and to be avoided. We argue that this perception of failure is very often incorrect.

9.3.1 What Failure Is Not

While considering what failure is, Cole (2011) suggests it is also helpful to consider, at least in broad terms, what failure is not. Drawing on the work of Maxwell (2000), Cole outlined four characteristics of what failure is not, which now are summarised:

- Failure is not always avoidable: All of us will fail at some time, probably more frequently than we succeed.
- Failure is not some 'freaky event': There is usually a process that leads to failure, such as not studying for a test, being careless, not realising there is a lack of clarity in our writing or not adequately understanding some concept or process.
- Failure is not always negative, even though it is very commonly seen to be: Some failures are the result of honest mistakes; these are not shameful. While we are conditioned to feel that mistakes are undesirable (e.g., when growing up, most individuals hear rhetorical questions/statements of the form "what's wrong with you?" or "get your act together"), not all mistakes are negative. Often we fail because we do not adequately know something or because we repeat past mistakes because we have not learned from them, or we deliberately continue to do the wrong thing. It is extraordinarily rare for anyone to hear the suggestion "it's OK to fail".
- Failure is seldom catastrophic: Assuming that a failure is not literally fatal, "every failure contains within it the seed of success – the opportunity to learn and improve" (Cole, 2011, p. 21). Adversity often leads to success, as is clear from many accounts of the development of new scientific knowledge or new technologies or other such advances. Although some failures are no more than just failures and so best avoided, other failures are valuable learning opportunities, if we choose to learn from them.

9.3.2 Why Failure Hurts

Humans generally do not like failure as it can highlight when we are incompetent, inaccurate, ineffective or 'wrong' (Firestein, 2015). Failing can give rise to a multitude of emotions, such as regret, guilt and shame (Cole, 2011), diminished perceptions of self (Conroy, 2003), avoidance and reduced risk-taking behaviours (Cetin et al., 2014). These emotions can in turn translate into reduced capacity to attempt new things, pursue study or make career choices (Simpson & Maltese, 2017). When failure leads to regret, we focus on past events rather than focusing on the future. Feelings of regret can affect our choices and actions, including diminishing our propensity to take risks as we are anchored to the past. If regret lingers it can turn into feelings of guilt, essentially the gap between how we behaved and how we feel we ought to have behaved (Cole, 2011). Guilt can then turn into shame, such as 'I am a bad person' or 'I am a failure'. These feelings are toxic as they can easily become linked to our identity, causing feelings of worthlessness and poor self-image. This can have flow on effects on behaviour, which can result in feelings of emptiness, withdrawal, loneliness and disempowerment (Cole, 2011). To deal with these emotions, Cole (2011) suggests acknowledging mistakes, taking responsibility for them, remedying any consequences (including forgiving ourselves and others, and apologising to others); then learning from the mistake can be a productive way of moving forward.

Aside from the personally confronting nature of failure, representations of failure in our schools and workplaces can also lead to failure aversion. Failure is often seen as undesirable, something to be avoided, seldom discussed, acknowledged and shared and often linked to our identity (Lottero-Perdue & Parry, 2017). Clark and Thompson (2013) suggest that part of what makes failure so undesirable is the lack of acknowledgment of the role of failure in our physical, written and oral communities. For example, in many research contexts, failure is notably absent from research articles, marketing websites and presentations. In a similar vein, Boutron et al. (2010) identified up to 40% of 'negative' research findings are communicated in a positive light or dismissed as anomalies in reports of the research, thus further highlighting our aversion to identifying failure.

Yet when failure is spoken about, interest and curiosity is often piqued as this resonates with our own sense of struggle. Hong and Lin-Siegler (2012) observed that humanising scientists by representing their struggles (as compared to just showcasing their achievements) increased students' perceptions of scientists as hardworking, fallible individuals who struggle. This change in perception led to improved student conceptual development, interest and ability to solve complex problems.

While failure often carries negative connotations (Simpson & Maltese, 2017), it has been agued as a central component of learning in a range of contexts (Kapur & Rummel, 2012). We now briefly note some of these before turning to the ways failure plays a role in the development of new knowledge in each of the disciplines of STEM.

9.4 Failure Is the Journey to Learning

"Courage allows the successful woman to fail – and to learn powerful lessons from failure – so that in the end, she didn't fail at all." Maya Angelou

It is sad that failure is often viewed as undesirable as it is through failures and mistakes that learning can abound (with this very often becoming possible only if one is willing to reflect on the failure and its causes). Success often represents an end point, the culmination of a journey of uncelebrated and underappreciated failures. Failures are an integral part of the learning process, which ultimately helps us become successful.

Clark and Thompson (2013) suggest four underappreciated aspects of failure:

- Failure reflects good academic practice; failure often occurs at the boundaries of innovation and progress.
- Failure is a teacher, helping to develop knowledge and skills, promoting personal growth and career progression.
- Failure drives progress, such as development of research, leading to unexpected avenues for inquiry and an opportunity to improve our work (Thomson & Kamler, 2012).
- Failure draws attention to injustice, drawing attention to unfair equalities which can "masquerade as ability, merit or luck" (Tessman, 2009). Failure can be a measure of discrimination.

The value of failure and its role in learning is generally becoming more recognised, and in what is often labelled the 'self-help literature' more and more popular. For example, Page (2011) talks about "failure chic", with a plethora of popular books devoted to the value of failure (examples include *How to be a successful failure* [Cole, 2011] and *Failing Forward: Turning Mistakes into Stepping Stones for Success*" [Maxwell, 2000]). Considerations of failure and its positive consequences are now also found in academic literature. For example, in April 2011 the Harvard Business Review published a 'failure issue' devoted to exploring the nature of failure (see https://hbr.org/archive-toc/BR1104).

9.5 Accounts of the Role of Failure in the Development of Science Knowledge

"Science has an inside secret: we fail all the time." Maryam Zaringhalam (2016)

When we look at the role and importance of failure specifically in science, we see a similar story to that presented above. Success is preferential to failure, as evidenced by the communication of successes over failures (there is no journal dedicated to reports of scientists' failures). The broad culture of science and the ways 'failure' is not portrayed are well represented by use of the term "secret" in the epithet by

Zaringhalam just above. Yet in the quest for new knowledge, failure in science is highly likely and is commonly asserted to be far more commonplace than success (e.g., Dreyfuss, 2019; Firestein, 2015; Parkes, 2019; Zaringhalam, 2016, 2017). Innovation is necessarily risky, and with risk comes the likelihood of failure. Yet a number of searches of *peer reviewed* science journals revealed an almost complete absence of any discussion about failure in accounts of science research. However, comments about the role and importance of failure in scientific endeavour were found to be abundant in blogs and other forms of informal (and so *not* peer reviewed and less constrained by long-term conventions) communication between research scientists.

Failure plays a large part in scientific endeavour. Yet research scientists claim that it is difficult to observe and appreciate the role of failure in the development of new knowledge due to the way science continues to be represented in formal publications, conference papers and applications for research funding. The long established and formalised processes for revealing new knowledge and other research conclusions very often report on completed science (and thus on science successes), rather than offering an authentic view of the actual processes of science that lead to the success (for elaboration about the lack of communication about failure in science, see Dreyfuss, 2019; Maestre, 2019; Zaringhalam, 2019). As Parkes (2019) suggests "comfortable science is an oxymoron. If we want to make new discoveries, that means taking a leap in the dark – a leap we might not take if we're too afraid to fail …. Science is high-stakes" (p. 5). Zaringhalam (2017) elaborates:

> Failure is the natural product of risk, and there's nothing riskier than the pursuit of ignorance—asking those big bold questions that probe the unknown. But while the practice of science is riddled with failures—from the banal failures of day-to-day life at the bench to the heroic, paradigm shifting failures that populate the book called Failure—many scientists are uncomfortable with the idea. We publish our innovations, the stories of how our ignorance led to success. Where the "publish or perish" mantra prevails, these stories are essential to making a name for ourselves and securing grant money. So there is little incentive to replicate the work of others or report experimental failure. In fact, there is barely a medium to publish these sorts of efforts, which are relegated to the bottom of the file drawer. (para. 9)

In this context of constant pressure to publish ('or perish') and apply for more funding, it is little wonder that most scientists are reluctant to spend time and effort communicating failures, or that many scientists drop out of the profession after a few years (Dreyfuss, 2019), or that people choose to *not* enter STEM professions in the first place (Simpson & Maltese, 2017).

Perception of failure can impact scientists' work, their propensity (better, lack of propensity) to communicate the role of failure in their work, and their longevity in the profession. Maestre (2019) highlights the role that anxiety also exerts on scientists, which can further limit communication of failure: "Focusing on success while living under continuous rejection may put more pressure on the work of our graduate students and postdocs, increasing their frustration and anxiety levels when their articles or applications are rejected" (p. 5). If we want scientists to acknowledge and speak about the important role of failure, this process needs to be valued in ways that de-stigmatise failure, and allow the forms of reward structures that are currently only accorded to refereed (and so conventional) publications.

As a consequence of scientists' aversion to being open about the role of failure in their research, a large proportion of all scientific endeavour goes unreported (Parkes, 2019) and a complete picture of science is therefore not possible (Zaringhalam, 2016). This not only has ramifications for the work of scientists but also presents a distorted view of the work of scientists to people outside science research. Importantly, these include most teachers of school science, as well as other groups such as lay people (as consumers of science) and individuals who might one day aspire to be scientists. Inaccurate and non-authentic portrayals of the scientific endeavour cannot give a full and accurate picture of how science is actually undertaken. The lack of recognition of the important and prevalent role of failure in science can impact students' choices to move into science as a career (Lin-Siegler et al., 2016; Whitlock, 2017). Stories of struggle can not only enhance the awareness of failure in science, but have also been reported to have positive effects on improving student motivation and performance in science, as students can then recognise science and the nature of scientific endeavour as being closer to their lived reality than they may have realised (Lin-Siegler et al., 2016).

A lack of understanding of the role of failure in the development and practice of science can also have consequences well beyond the specific realms of science and science education (and in the extreme, consequences in this century that were largely unimaginable even just 50 years ago). The extended quote from Zaringhalam (2017) we have given just above begins with "failure is the natural product of risk" (para. 9). One widespread use of science that must always involve 'risk' (and so 'failure') is forecasting – of the weather to be expected in the next week, of the future consequences of an individual's medical condition, of the magnitude of global temperature increases that will result from future global emissions, and so on. Forecasting is necessarily a 'probabilistic' exercise, and cannot be absolute. The combination of a lack of lay acceptance (and understanding) of the necessarily probabilistic nature of (and so 'risk' in) forecasting in science-related matters and the increasingly litigious nature of democratic societies has now led to what many see as extreme consequences. These are well illustrated by the case of the earthquake in 2009 that caused major damage to the city of L'Aquila in central Italy, and multiple human deaths. L'Aquila is in a region of ongoing seismic activity. Six seismologists had attempted to forecast the then current level of seismic risk to the city and its inhabitants shortly before the 2009 earthquake hit, with, for example, one forecast assessing (forecasting) matters as 'normal' and indicating that inhabitants of this region should remain in the region. Subsequently the six, and the one government official responsible for sending the seismologists to the region to undertake the forecasting, were convicted of multiple manslaughters and sentenced "to six years jail for having given false assurances to the public before an earthquake hit...L'Aquila" (Davies, 2012, para. 2). This was despite the undisputed fact that the current capacity of science to forecast earthquakes resulting from earth movement along a specific fault is no more than probabilistic, and cannot include any specific timeframe.[2] (Appeals resulted in the overturning of the convictions for the seismologists in 2014 and for the government official in 2016.)

Failure in science is also evident in standard accounts of knowledge development through stories of serendipity, happenstance and blunders (Livio, 2013), such as Fleming discovering penicillin in unwashed petri dishes (although such stories are almost always represented in such simplified ways as to indicate that it is serendipity that is central and not failure). Accounts of science consistently fail to communicate how "fortune favours the *prepared* mind" (Louis Pasteur, emphasis added).

Another consequence of scientists' unwillingness to be open about their failures (Dreyfuss, 2019) is the hindrance of the progress of science (Madisch, 2017). If scientists do not share their failures (or even hold stronger positions about failure such as 'failures are something shameful') then much scientific endeavour will consist of replication of previously experienced failures. Parkes (2019) further asserts that knowing about failures would help speed up scientific progress. Initiatives like open access publishing and being more open about failures could help normalise the role of failure. Scientists have attempted to do this through initiatives such as the free access website F1000 research (see https://f1000research.com/), where scientists publish negative and null data results, or a CV of failures, as a Princeton University Professor of psychology and public affairs, Professor Johannes Haushofer did when he wrote a "CV of failures" (see https://www.princeton.edu/~joha/Johannes_Haushofer_CV_of_Failures.pdf). While he had intended this to be for his students, it surprisingly went viral (Swanson, 2016). The important role of failure has also led to the development of the interdisciplinary Education for Persistence and Innovation Centre (EPIC) at Teachers College Columbia University, led by Professor Xiaodong Lin-Siegler. The core purpose of this centre is to study the critical role that failure plays in innovation, learning, leadership and career progression (see http://epic.tc.columbia.edu/).

9.6 Accounts of the Role of Failure in the Development of Technology and Engineering Knowledge

> "Don't read success stories, you will only get a message. Read failure stories, you will get some ideas to get success." Abdul Kalam

Kalam, the author of the quote above, was an aerospace engineer of renown (and later President of India). The point made in the quote is in stark contrast to accounts of failure in research in science, the role of failure is widely accepted and celebrated in the development of new knowledge in technology and engineering. Thomas

[2] This is well illustrated by reference to one of the most well-known fault lines, the San Andreas Fault, which basically runs down the coastline of California. It is recognised by both seismologists and non-scientists in California that it is highly likely that there will be another earthquake as catastrophic as that in San Francisco, 1908 or Los Angeles, 1857—but this may be next year or next century, and it may be anywhere along the Fault.

Edison, a technologist/engineer researcher, not a research scientist, is claimed to have said, "I have not failed. I've just found 10,000 ways that won't work". This quote further reflects the very different broad culture with respect to 'failure' in Technology and Engineering (TE). TE is driven by innovation and is responsible for the multitude of human-made objects and structures that exist to enhance our lives, yet the principles and processes of TE are seldom understood (Petroski, 1982). Often, TE is used to create a design solution for a problem, but as Petroski (2006) elaborated, the innovation and development of new technologies can also follow from the failures of existing technologies to perform as we hoped or as promised. Failure therefore plays a role in the development of new TE innovations, as testing something new to ensure it is fit for purpose and is safe for use necessarily requires trial and error—rigorous, systematic and controlled, it is true, but nevertheless an organised form of 'trial and error'. Innovation necessarily involves failure (Engel, 2018) and failure-tolerant environments are known to nurture innovation (Townsend, 2010). In TE, failure also acts as a way of identifying areas for improvement in innovative products and structural designs.

In science, failure is experienced but seldom spoken about; in sharp contrast, failure is at the very heart and soul of technology and engineering. Common professional phrases like 'tested or engineered to failure' position mistakes (failure) and learning from failure as essential parts of the design process, and accounts of this for specific design cases are present in relevant literature (e.g., Gomoll et al., 2018). Technology and engineering draw on similar design processes to develop new initiatives, designs and innovations. Although multiple models for engineering and design thinking have been proposed, similarities in the design process are evident across these models, with testing, failure and retesting being integral aspects of the models. A literature search identified eight such detailed models (see Table 9.2). It seems likely to us that the level of detail in each of these models is a consequence of the motivation for creating each model: to guide curriculum or learning development relevant to design (for schools/undergraduate in eight cases, and for graduate/professional learning in the ninth).

There are some significant features common to many of the models in Table 9.2. All but one start by identifying and describing a real-world problem or need or issue that is to be the focus of the design task; the exception, the Stanford model, positions "empathising" (understanding people within the context of the design task) as the very first step. In each model the initial outcome of the first step is uncertain. So, the next step is to become more informed about the problem/need/issue through ideation, imagining, researching the problem, etc. Some of the models recognise that this step can be a non-linear process, where thinking can jump back and forth between stages, such as moving between empathising, identifying a need or problem, and generating knowledge about the problem as these processes are undertaken at the same time.

The next step is to generate a prototype or model, or propose other appropriate forms of possible solutions to the problem, etc. This is then followed by a testing and improvement phase, where success and failure are used as measures of appropriateness of the design. This in turn is followed by an improvement phase, where

Table 9.2 Examples of engineering and design process models

Massachusetts Sci & Tech/ Engineering Curriculum Framework *(Cyclical and stepwise)*	Engineering Is Elementary (EiE) *(Cyclical)*	The new "BEST" model *(Stepwise, cyclical, interactions between steps)*	National Center for Engineering and Tech Ed (NCETE) *(Stepwise with cycles within the process)*	The UTeach Engineering project *(Linear, cycle in middle)*	Next Generation Sci Standards (NGSS) *(Three interacting stages)*	Stanford Model for Design Thinking *(Stepwise)*	PictureSTEM *(Two stages with feedback steps)*	Systems engineering model [Engel] *(Stepwise with feedback)*
						Empathize		
Identify the need or problem	Ask	Ask	Identify need or problem	Identify the need	Define	Define	Problem: Define	Development: Definition
Research the need or problem	Imagine	Imagine	Research need or problem	Describe: Describe the needs and characterise and analyse the system		Ideate	Problem: Learn	Design
Develop possible solutions	Plan	Cycle: Plan	Develop possible solutions	Generate: Generate concepts	Develop solutions		Solution: Plan	Implementation
Select best possible solution			Select best possible solution	Generate: Select a concept				
Construct a prototype	Create	Cycle: Create	Construct a prototype	Embody: Embody the concept		Prototype	Solution: Try	Integration
Test and evaluate the solution		Cycle: Test	Test and evaluate solution	Embody: Test and evaluate the concept		Test	Solution: Test	Qualification
		Cycle: Improve						

Massachusetts Sci & Tech/ Engineering Curriculum Framework *(Cyclical and stepwise)*	Engineering Is Elementary (EiE) *(Cyclical)*	The new "BEST" model *(Stepwise, cyclical, interactions between steps)*	National Center for Engineering and Tech Ed (NCETE) *(Stepwise with cycles within the process)*	The UTeach Engineering project *(Linear, cycle in middle)*	Next Generation Sci Standards (NGSS) *(Three interacting stages)*	Stanford Model for Design Thinking *(Stepwise)*	PictureSTEM *(Two stages with feedback steps)*	Systems engineering model [Engel] *(Stepwise with feedback)*
Communicate the solution(s)		Share	Communicate solution				Communication/ teamwork	
Redesign	Improve		Redesign	Embody: Refine the concept	Optimize			
			Finalize design	Finalise and share the design AND evolve the design			Solution: Decide	
								Post development: Production, use and disposal

changes and redesign, based on data obtained during testing of the model, take place. Each model finishes with some kind of communication phase, where the design is shared with others. The Systems engineering model (Table 9.2) goes a step further and considers issues related to the eventual disposal of the final product of the design process after the end of its functional life (we infer this to be a consequence of the very different and much more educationally advanced target group for which this model has been created).

In each model, the process requires a degree of creative thinking (see Ellerton and Kelly, Chap. 2, this volume), such as fluency and flexibility of thinking (especially in ideation), originality to come up with new ideas, capacity to redefine and replace existing ideas, and a willingness to accept uncertainty. The process also requires critical thinking to make judgements about alternative prototypes and possible solutions.

9.6.1 Role of Failure in the Design Process

"Failure is the key to success." Michelle Obama

In a manner that is different to science, the work of engineers and technologists is heavily scrutinised and subjected to public testing. Testing of an object or structure ensures the design is fit for purpose and will work as intended. People can observe the fruits of an engineer's labours, but can also suffer the consequences if the design fails, such as when bridges collapse or when a car does not start. The consequences of TE errors are very often far more obvious than for other professions, such as scientists, mathematicians, lawyers or accountants (Engel, 2018). There is also a difference in the design and development of objects always intended to be produced for mass production, as compared with those objects that are seen to be unique throughout the design process. Mass-produced objects often undergo further debugging and evolution after they are released to the public (Petroski, 1982). On the other hand, larger civil engineering structures that are effectively unique, such as a single bridge or a single building, need to be fit for purpose from the first stages of construction. Learning from failure (so as to productively build on previous failures) plays a pivotal role in any design process, and especially in examples such as large-scale engineering projects which cannot be tested and consequently modified in design as they are built.

9.6.2 Failure Analysis Methods

"It's the failure that leads to success, while prolonged success leads to failure." Henry Petroski

The role of failure as a learning tool in TE can be seen through engineering failure investigations. There are many different investigation types, such as commercial (insurance claims and contractual disputes), liability (to establish fault), accident (what happened and who was to blame) and research (generic improvements and improving understanding) (Matthews, 1998). The premise of failure analysis case studies is to critically analyse the nature of a TE failure and thus to publicly offer a way for engineers (or other designers) to examine, discuss, and share (including via publishing) detailed analyses, with the intent of avoiding similar incidents and improving future related design. For example, with engineering equipment, which usually has a mechanical basis, failure can take the form of component fatigue (Matthews, 1998). Understanding the nature of failures of various materials offers insights about future materials selection, and so be a better fit for a particular purpose.

The articulation of failure cases is diverse. It is found in books (see for example Jones, 1998) and dedicated journals, such as *Engineering Failure Analysis,* which accepts papers that describe "the analysis of engineering failures and related studies", and *Case Studies in Engineering Failure Analysis*, a journal whose title makes clear the nature of the papers it seeks. Dissemination of TE failures also frequently occurs through conferences (a mode of dissemination that is of greater prominence and significance in engineering than in many other fields, in part because forms of refereed conference proceedings are common), such as the *International Conference on Engineering Failure Analysis*. Medicine has a failure investigation process in the form of Morbidity and Mortality conferences (MMC) to analyse adverse events, errors and shortcomings in patient care and treatment (Bal et al., 2014). Sometimes analysis of failure is very public, as with the 1983 Challenger space shuttle disaster (Rogers Commission Report, see https://history.nasa.gov/rogersrep/genindex.htm).

As well as drawing on failure to ensure objects are fit for purpose, TE also involves engineering objects to fail in predictable ways to ensure their safety and continued usefulness. Petroski (1997) explains:

> We actually want certain things to fail and break, for otherwise we would be frustrated in their use and possibly even harmed by their existence. The challenge to the engineer in this case is to design systems and devices that have well-defined and predictable failure and breaking points so that such physical phenomena as collapse or fracture happen in the way and at the time they are supposed to. (p. 412)

Purposeful or built-in failure mechanisms include things such as fuses, pressure valves, cracks and purposeful gaps in bridges and pavements. Used in this way, failures are designed into objects to act as a 'fail safe' to ensure products and structures are engineered for maximum usefulness and safety. Failure in this instance has different ramifications compared to science, where failure is part of the process that leads to success, rather than something that is purposefully worked into the design.

Failure in TE also has a different value attached to it than does failure in science. Petroski (1997) argues this via the metaphorical use of the example of peeling an apple with a knife, where the intended purpose of removing the peel is "to cause the failure of the skin to continue to adhere to the fleshy part of the apple" (p. 413).

Used in this way, 'failure' is used to explain how the system has changed, in this case, by causing a failure that is desirable to some (those who prefer apples to be eaten skinless). From this metaphor, Petroski points to the ways in which failure at one point in a design or problem solving process is recognized as often being a central step forward in the eventual completion of the design or crafting of a solution to a problem.

9.7 Accounts of the Role of Failure in the Development of Mathematics Knowledge

Early in this chapter we quoted William Whewell's succinct statement about failure and success: "Every failure is a stem to success". Whewell was a remarkable nineteenth century polymath who made particularly important contributions to new knowledge in mathematics, philosophy, and the nature and forms of the processes of development of new ideas in science – to use Whewell's terms, 'scientific method'. This included the then definitive account of the nature of induction and the logic of discovery, although his work was certainly not confined to these matters. We turn to Whewell again here in order to point to the very different nature of knowledge in what is widely regarded as a system of logic such as mathematics, when compared with empirically based disciplines (S, T, E) with which he was concerned when writing 'every failure is a stem to success'. Whewell (as reproduced in Butts, 1968), writing in 1837, saw the certainty of mathematics as arising from its being founded on *axioms* (emphasis in original), and conducted by steps that can each, if required, be stated as *syllogisms*. The certainty and conclusiveness of axioms and syllogisms in turn rests on initial *definitions.* (The conclusion that mathematics rested on definitions was also reached by other philosophers and mathematicians in this period.) This view of the nature of mathematics knowledge is consistent with the school of philosophy that is known as 'Symbolic Logic' (Shapiro & Kouri Kissel, 2018).

This view that mathematics as knowledge is derived via logic from a set of initial definitions is still widely accepted today, but no longer universally. The growth in alternative perspectives on the ways mathematics knowledge is created, and the surprise with which alternatives are greeted by some, are succinctly described by Devlin (2008):

> Recent years have seen a growing acknowledgement within the mathematical community that mathematics is cognitively/socially constructed. Yet to anyone doing mathematics, it seems totally objective. (p. 359)

In recent times, ideas derived from the social construction of knowledge that lead to this less certain view of the nature of mathematics have become more common.[3]

[3] See, for example, the work of the radical constructivist Ernst von Glasersfeld (e.g. von Glasersfeld, 1995), and the work on philosophy of both mathematics and mathematics education by Paul Ernest (e.g. 1997).

And this century is seeing the emergence of more and diverse new thinking about the nature of mathematical knowledge, including discussions of the epistemology of mathematics that in previous times were at best extremely rare. For example, the May 2008 issue of the analytic philosophy journal *Erkenntnis* is devoted to the theme 'Towards a new epistemology of mathematics'. Even inductive processes have been used in forms of developing mathematical knowledge, most obviously with Fermat's last theorem[4] which was for many years accepted as true solely on the basis of induction from the correctness of the theorem for many specific cases.

Whether one sees the nature of mathematics to be purely logical, and so of the traditional and more widely held view, or to be a socially constructed form, or to be something different again, is not the central issue here. What is critical is that our literature searching has not found any mention of 'failure' as an issue in the development of mathematical knowledge. Although we certainly acknowledge that accounts of the development of mathematical proof as 'empirical' or 'experimental' (e.g., Baker, 2008; Buldt et al., 2008) can be taken to imply a role for failure, the term is not used.) Further, 'failure' is not mentioned at all in iconic texts concerned with the nature of mathematics, such as Courant and Robbins (1961); nor does it appear at any point in the comprehensive (almost 2500 page) anthology of a millennium of the literature of mathematics created by Newman (1956). Indeed, as Burton (2001) observed, there is "surprisingly little to be found which critically assesse[s] the epistemology of mathematics" (p. 589). More significantly, Burton advanced this observation early in her report of a detailed and intensive study of 35 research mathematicians and their approaches to their own learning of and developing of new knowledge about mathematics. At no point was 'failure' raised by any of the 35 participants.

It is clear that 'failure' does not play even a minor role in accounts of the development of new knowledge in the way of knowing that is mathematics. We assert that the differences between 'M' and 'S/T/E' with respect to the ways 'failure' is represented in accounts of development could hardly be greater.

We also note that it is a completely different matter when one considers the *learning* of mathematics. In general, in the hands of a skilled teacher whose focus is on conceptual learning of mathematics, 'failure' has powerful potential for enhancing this learning. More specifically, two relevant constructs have been explored in studies of mathematics learning: "fear of failure" (sometimes described as "mathematics anxiety", Foley et al., 2017), and, less prominently, "fear of success" (that is, fear of the consequences of success in mathematics learning, something that has been one contributor to the gender differences in participation in mathematics courses; e.g., Leder, 1982).

[4] Fermat's last theorem (that no three positive integers a, b, and c satisfy the equation $a^n + b^n = c^n$ for any integer value of n greater than 2) was stated by Pierre de Fermat in 1637, together with an assertion that he had proven this but without giving the proof. A proof advanced in 1994 has become accepted.

9.8 The Role of Failures in School Learning of the STEM Disciplines

"Anyone who has never made a mistake has never tried anything new." Albert Einstein

We have argued that there is misalignment between each of the individual STEM disciplines in terms of the ways that the role of failure in the development of new knowledge is represented in each discipline. We noted in the Introduction to this chapter that the way failure is represented in each discipline *per se* impacts on the ways that discipline is represented to and perceived by those outside the discipline. This is most obviously the case in the education of students in each of the disciplines, and in integrated STEM. That is, there is also significant misalignment in primary and secondary schooling contexts in the ways the individual STEM disciplines portray failure and its role, a misalignment which has significant impact on the learning of students. For example, it is difficult to celebrate the central role of failures in the development of the discipline of science when the nature of school science so often emphasises certainty of knowledge (the curriculum as a 'rhetoric of conclusions'). This certainty of knowledge is reinforced when 'recipe style' laboratory tasks with pre-determined steps and outcomes are used, and when assessments treat science as a rigid body of facts to be learned and regurgitated. Technology education, in a number of countries, attempts to link to real world and authentic contexts, such as food and fibre production, but whether or not teachers and schools enable students to experience the nature of the 'design process', including the beneficial consequences that can emerge from 'failure', is dependent on a wide range of contextual factors (e.g. curriculum, school, teacher).

More authentic experiences with STEM disciplines can help students recognise the value and role of failure in both real-world STEM contexts and in their own learning of STEM, as such experiences may lead to greater learner awareness of the value and prevalence of failure in the development and processes of these disciplines. We now list some ways this might be achieved.

- Sharing stories of actual experiences of scientists, most importantly including how they struggled intellectually and personally and how they actually made their 'discoveries' over a period of time and through a range of experiences (usually involving struggle, failure and/or serendipity, with perhaps controlled scientific investigation having some part) (Lin-Siegler et al., 2016). Such 'struggle stories' help students to feel more connected to scientists and enable students to see themselves as not being too dissimilar to the scientists, which in turn can impact whether students choose to select STEM disciplines as a future career.
- Discussing the nature of success and failure – acknowledging that bad processes do not always lead to success and correct processes might still result in failed outcomes (Dahlin et al., 2018).
- Defining what success and failure can look like (McGrath, 2011), and considering how these are similar and different across the separate disciplines of STEM.

- Ensuring failure is efficient, forward focused and cost effective (a perspective already common in technology and engineering, but much less so in science and mathematics).
- Focusing on process and journey rather than an end product; having conversations about the nature of failure, considering what works and what does not work and why (McGrath, 2011).
- Focusing on building capabilities and dispositions which handle failure, such as resilience, adaptability, critical and creative thinking, and collaboration.

Although it is beyond the scope of this chapter, we most certainly acknowledge the profoundly important role of failure in the processes of learning, whether it is about the epistemologies of a discipline or the concepts and relationships of the discipline (e.g., Searle et al., 2018; Zieglar & Kapur, 2018) or the development of the capacity to be creative or to think critically (see Ellerton & Kelly, Chap. 2, this volume). We have already noted aspects of this at the end of the brief section about failure and any role of this in the development of the discipline knowledge of mathematics. Indeed, the many critiques that exist of 'conventional', stereotypical school mathematics classrooms could be recast in terms of the complete lack of attempts to use a student's 'failure' in tackling a specific problem as a path to developing that student's understanding, something emphasised by the construct "fear of failure" in mathematics learning. Classroom environments that fail to recognise, discuss and share the value and necessity of failure in learning can stifle learning (Dahlin et al., 2018), in ways we see as broadly consistent with the negative impact on the development of the disciplines of S, T, E and M when failure is ignored or if there is a pretence that failure does not occur. We have argued in this chapter that failure is a critical component of success in the STEM disciplines. However it is rarely recognised as such in formal accounts of the processes of some of these disciplines. In conclusion, we note that this has clear lessons for STEM classrooms, lessons that include the list of dot points above. These dot points provide ways of beginning to think about introducing the notion of failure into STEM education, and so recognising the importance of failure in each of the disciplines *and* in learning. In addition, we would add the following points to the list above:

- Develop a culture that values and openly acknowledges failure, is forgiving and celebrates failure as being part of the learning process (McGrath, 2011)
- Discuss the nature of success and failure – acknowledge that bad processes do not always lead to success and correct processes might still result in failed outcomes (Dahlin et al., 2018)
- Nurture a growth mindset which focuses on what needs improving rather than what failed
- Reflect on and articulate learning – individually and as a group (Townsend, 2010)

References

Baker, A. (2008). Experimental mathematics. *Erkenntnis, 68*, 331–334.

Bal, D. G., Sellier, D. E., Tchouda, D. S., & François, D. P. (2014). Improving quality of care and patient safety through morbidity and mortality conferences. *Journal for Healthcare Quality, 36*(1), 29–36. https://doi.org/10.1111/j.1945-1474.2011.00203.x

Boutron, I., Dutton, S., Ravaud, P., & Altman, D. G. (2010). Reporting and interpretation of randomized controlled trials with statistically nonsignificant results for primary outcomes. *The Journal of the American Medical Association, 303*(20), 2058–2064. https://doi.org/10.1001/jama.2010.651

Buldt, B., Löwe, B., & Müller, T. (2008). Towards a new epistemology of mathematics. *Erkenntnis, 68*, 309–329.

Burton, L. (2001). Research mathematicians as learners: And what mathematics education can learn from them. *British Educational Research Journal, 27*, 589–599.

Butts, R. E. (Ed.). (1968). *William Whewell's theory of scientific method*. University of Pittsburgh.

Cetin, B., Ilhan, M., & Yilmaz, F. (2014). An investigation of the relationship between the fear of receiving negative criticism and of taking academic risk through canonical correlation analysis. *Educational Sciences: Theory & Practice, 14*(1), 146–158.

Clark, A. M., & Thompson, D. R. (2013). Successful failure: Good for the self and science. *Journal of Advanced Nursing, 69*(10), 2145–2147. https://doi.org/10.1111/jan.12125

Cole, K. (2011). *How to be a successful failure: A practical guide to messing up big time, the right way*. Xlibris.

Conroy, D. E. (2003). Representational models associated with fear of failure in adolescents and young adults. *Journal of Personality, 71*(5), 757–784. https://doi.org/10.1111/1467-6494.7105003

Courant, R., & Robbins, H. (1961). *What is mathematics?* Oxford University.

Dahlin, K., Chuang, Y., & Roulet, T. (2018). Opportunity, motivation, and ability to learn from failures and errors: Review, synthesis, and ways to move forward. *Academy of Management Annals, 12*(1), 252–277. https://doi.org/10.5465/annals.2016.0049

Davies, L. (2012, October 2). *Jailing of Italian seismologists leaves scientific community in shock*. The Guardian: Australia Edition. Retrieved from https://www.theguardian.com/world/2012/oct/23/jailing-italian-seismologists-scientific-community

Devlin, K. (2008). A mathematician reflects on the useful and reliable illusion of reality in mathematics. *Erkenntnis, 68*, 359–379.

Dewey, J., Boydston, J.-A., & Rorty, R. (2008). *The later works of John Dewey, 1925–1953*. Southern Illinois University.

Dreyfuss, E. (2019). Scientists need to talk more about failure. *Wired*. Retrieved from https://www.wired.com/story/scientists-need-more-failure-talk/

Engel, A. (2018). *Practical creativity and innovation in systems engineering*. Wiley.

Ernest, P. (1997). *Social constructivism as a philosophy of mathematics*. SUNY.

Failure. (n.d.). T*he Merriam-Webster.com Dictionary*. Retrieved from https://www.merriam-webster.com/dictionary/failure

Firestein, S. (2015). *Failure: Why science is so successful*. Oxford University.

Foley, A. E., Herts, J. B., Borgonovi, F., Guerriero, S., Levine, S. C., & Beilock, S. L. (2017). The math anxiety-performance link: A global phenomenon. *Current Directions in Psychological Science, 26*, 52–58.

Gomoll, A., Tolar, E., Hmelo-Silver, E., & Šabanović, S. (2018). Designing human-centered robots: The role of constructive failure. *Thinking Skills and Creativity, 30*, 90–102.

Hong, H.-Y., & Lin-Siegler, X. (2012). How learning about scientists' struggles influences students' interest and learning in physics. *Journal of Educational Psychology, 104*(2), 469–484. https://doi.org/10.1037/a0026224

Jones, D. R. H. (1998). *Failure analysis case studies: A sourcebook of case studies selected from the pages of engineering failure analysis 1994–1996*. Elsevier.

Kapur, M., & Rummel, N. (2012). Productive failure in learning from generation and invention activities. *Instructional Science, 40*(4), 645–650. https://doi.org/10.1007/s11251-012-9235-4
Leder, G. C. (1982). Mathematics achievement and fear of success. *Journal for Research in Mathematics Education, 13*, 124–135.
Lin-Siegler, X., Ahn, J. N., Chen, J., Fang, F.-F. A., & Luna-Lucero, M. (2016). Even Einstein struggled: Effects of learning about great scientists' struggles on high school students' motivation to learn science. *Journal of Educational Psychology, 108*(3), 314–328. https://doi.org/10.1037/edu0000092
Livio, M. (2013). *Brilliant blunders*. Simon & Schuster.
Lottero-Perdue, P. S., & Parry, E. A. (2017). Perspectives on failure in the classroom by elementary teachers new to teaching engineering. *Journal of Pre-College Engineering Education Research (J-PEER), 7*(1), 1–24.
Madisch, I. (2017). Why scientists must share their failures. *Scientific American*. Retrieved from https://blogs.scientificamerican.com/guest-blog/why-scientists-must-share-their-failures/
Maestre, F. T. (2019). Ten simple rules towards healthier research labs. *PLoS Computational Biology, 15*(4), e1006914. https://doi.org/10.1371/journal.pcbi.1006914
Manalo, E., & Kapur, M. (2018). The role of failure in promoting thinking skills and creativity: New findings and insights about how failure can be beneficial for learning. *Thinking Skills and Creativity, 30*, 1–6.
Matthews, C. (1998). *A practical guide to engineering failure investigation*. Professional Engineering.
Maxwell, J. C. (2000). *Failing forward: How to make the most of your mistakes*. Thomas Nelson.
McGrath, R. G. (2011). Failing by design. *Harvard Business Review, 89*(4), 76–83.
Newman, J. R. (Ed.). (1956). *The world of mathematics (volumes 1–4)*. Simon & Schuster.
Page, C. (2011, June 26). Failure is now praised as leading to success. *Philadelphia Tribune*, p. 1. Retrieved from https://search-proquest-com.ezproxy.lib.monash.edu.au/docview/877333579?accountid=12528&rfr_id=info%3Axri%2Fsid%3Aprimo
Parkes, E. (2019, January 10). Scientific progress is built on failure. *Nature*. Retrieved from https://www.nature.com/articles/d41586-019-00107-y
Petroski, H. (1982). *To engineer is human: The role of failure in successful design*. Macmillan London Limited.
Petroski, H. (1997). Designed to fail. *American Scientist, 85*(5), 412–416.
Petroski, H. (2006). *Success through failure: The paradox of design*. Princeton University.
Searle, K., Litts, B., & Kafai, Y. (2018). Debugging open-ended designs: High school students' perceptions of failure and success in an electronic textiles design activity. *Thinking Skills and Creativity, 30*, 125–134.
Shapiro, S., & Kouri Kissel, T. (2018). Classical logic. In E. N. Zalta (Ed.), *The Stanford encyclopedia of philosophy* (Spring 2018 Edition). Retrieved from https://plato.stanford.edu/archives/spr2018/entries/logic-classical/
Simpson, A., & Maltese, A. (2017). "Failure is a major component of learning anything": The role of failure in the development of STEM professionals. *Journal of Science Education and Technology, 26*(2), 223–237. https://doi.org/10.1007/s10956-016-9674-9
Success. (n.d.). *The Merriam-Webster.com Dictionary*. Retrieved from https://www.merriam-webster.com/dictionary/success
Swanson, A. (2016, April 28). Why it feels so good to read about this Princeton professor's failures. *The Washington Post*. Retrieved from https://www.washingtonpost.com/news/wonk/wp/2016/04/28/it-feels-really-good-to-read-about-this-princeton-professors-failures/
Tessman, L. (2009). Expecting bad luck. *Hypatia, 24*(1), 9–28.
Thomson, P., & Kamler, B. (2012). *Writing for peer reviewed journals: Strategies for getting published*. Routledge.
Townsend, W. R. (2010). Innovation and the value of failure. *International Journal of Management and Marketing Research, 3*(1), 75–84.
von Glasersfeld, E. (1995). *Radical constructivism: A way of knowing and learning*. Falmer.

Whitlock, S. (2017). *One reason young people don't go into science? We don't fail well.* Retrieved from https://www.statnews.com/2017/03/21/science-research-fail/

Zaringhalam, M. (2016). *Failure in science is frequent and inevitable – And we should talk more about it.* Retrieved from https://blogs.scientificamerican.com/guest-blog/failure-in-science-is-frequent-and-inevitable-and-we-should-talk-more-about-it/

Zaringhalam, M. (2017). *An experiment that didn't work: My PhD thesis research was a dead end, but that's why it was important.* Retrieved from https://blogs.scientificamerican.com/observations/an-experiment-that-didnt-work/?print=true

Zaringhalam, M. (2019). *To groom better scientists, harness the power of narrative.* Retrieved from https://undark.org/2019/01/24/to-groom-better-scientists-harness-the-power-of-narrative/

Zieglar, E., & Kapur, M. (2018). The interplay of creativity, failure and learning in generating algebra problems. *Thinking Skills and Creativity, 30*, 64–75.

Chapter 10
Humanistic Goals for Science Education: STEM as an Opportunity to Reconsider Goals for Education

Michael Tan

Abstract In response to economic rationales for STEM education, I propose here instead that science educators should consider the broader perspective if we desire that STEM education be educative. While economic rationales are important, they do not sufficiently attend to certain conditions of our time that science and technology may exacerbate. Specifically, I discuss reductionism as a key method of the natural sciences, the risk of opacity of technological function, and the hidden politics of technological systems. I argue that STEM educators, now enabled to attend to interdisciplinary concerns, have the opportunity to respond to these problems, and I propose a candidate pedagogical orientation for this purpose. We can no longer carry on 'business as usual', and we need new ideas, new narratives, and new ways forward. This chapter is an attempt to think in such terms.

Keywords STEM curriculum · Bernard Stiegler · Hannah Arendt · Humanism · Goals of STEM education

10.1 Introduction

STEM education is a recent initiative that has widespread influence around the world. While many stakeholders will associate STEM with engaging lessons filled with interesting devices that promise to motivate student learning, I argue here that we should instead take this moment as an opportunity to reconsider some goals for education. The concern here is that we simply transfer old wine into new bottles, and do not sufficiently attend to several problems to which current approaches to STEM instruction may contribute. Instead, I suggest that STEM educators should consider the characteristic humanness embedded in the pursuit of mass public education, and explicitly embrace the risk inherent in attempts to interfere in the lives

M. Tan (✉)
National Institute of Education, Nanyang Technological University, Singapore, Singapore
e-mail: michael.tan@nie.edu.sg

A. Berry et al. (eds.), *Education in the 21st Century*,
https://doi.org/10.1007/978-3-030-85300-6_10

of others. In essence, the successes of STEM in manipulating things may implicitly convince us to adopt a similar perspective in our (albeit well intended) manipulation of people. I also want to argue that these successes are not without costs, and that these costs for education may exceed what we should be prepared to pay. Given the panoply of possible disasters that await 'business as usual', we desperately need new ways of being. STEM education can provide such an opportunity, but only if we are prepared to reconsider our goals for education and embrace the risk of creating the unforeseeable.

10.2 What Is STEM Education?

Science, Technology, Engineering, and Mathematics (STEM) education has been a worldwide movement, with much funding from state and national governments, and rapidly expanding attention from education researchers. Despite these actions, STEM education has largely been ill defined; for example a recent review (Martín-Páez et al., 2019) indicates a certain opacity to the definition of STEM. In the early 1990s the US National Science Foundation began to use the acronym 'SMET' to refer to school and undergraduate Science, Mathematics, Engineering, Technology and the potential of these for matters such as the future careers of students. The acronym was changed to STEM by NSF in the early 2000s, and the term quickly began to take a life of its own and gain significance when it was recognised by multiple agencies that an education in the STEM disciplines was important for economic competitiveness. Soon after, STEM education arose in response to match the growing human capital demand.

While well intentioned, the curriculum intentions of STEM have been hard to translate into the structures and traditions of schooling. Grounded in the essential unity of phenomena, the STEM curriculum movement has encountered schools more accustomed to traditional disciplinary boundaries: phenomena seldom present themselves solely in terms of the disciplinary traditions into which we have organised universities and schools. This has led to a situation where school subjects present "a disconnected and inconsistent variety of skills and facts" (Martin-Páez et al., 2019, p. 802). In considering an approach to instruction, Martín-Páez et al. (2019) find that engineering is a useful 'hinge' discipline for STEM, due to the employment of robotics, engineering design, and engineering-based problems. Indeed, the three features of: (i) high contextual integration, (ii) realistic problem solving; and (iii) construction of proposed solutions, appear as strong emblematic STEM features. Often found in commentary accompanying these features are authors' assertions of the changed socio-economic conditions, with frequent reference to '21st century' competencies, or work conditions that require more than competencies involving abstract representations. For an example of the latter, we are wont to remind readers that the world has changed, that open-ended, ill-structured, and interdisciplinary project based problem solving are new skills that are needed going forward (e.g., Capraro et al., 2013).

A strong practice-based argument is also detected by Tytler et al. (2019): STEM can be interpreted as an attempt to make school science more closely resemble science in practice. However, as one might expect from attempts to integrate topics that have been strongly differentiated from one another, such integrations are unlikely to please all parties. Tytler et al. (2019) cite both Clarke (2014), who questions the ability of STEM to support significant disciplinary learning in mathematics or science, and Lehrer (2016), who likens STEM to an 'epistemic stew'. Significantly, a recurrent critique has been of the apparent ability of STEM to engage students, most notably with the use of 'fun' activities. Such lessons, upon further investigation, often reveal less than positive results, leading some researchers to ask: "it looks like fun, but what are they learning?" (Bevan et al., 2015).

In this chapter, I will consider the fundamental curriculum question 'what is it that students ought to be learning, and why?' specifically in the context of STEM. To be sure, there will always be gaps between the intended and the implemented curriculum, and there will be a wide range of reasons for this. However, the aim of this chapter is not to take the intended curriculum for granted, and then query the nature of the gap, or propose methods for its closure. Given that it is most frequently the case that national and state offices determine the curriculum for public schooling, and where this is not the case textbooks are often the *defacto* determinate of the curriculum, it can feel futile for educators seeking to shift the agenda for education in general, and STEM education in particular. However, I want to suggest here that the mechanism of the "discursive gap" (Moore & Muller, 2002) allows us some latitude in deciding what it is that should be implemented. In simple terms, this gap refers to the fact that there will always be a gap between a phenomenon and what can be said about it. The theoretical terms that one uses to describe just what it is that is happening will only ever describe limited aspects of the phenomenon, and hence there will always be room for interpreting curriculum goals. It is not my intention to illustrate the use of the concept of the discursive gap in this chapter, although there certainly is room for such research. Instead, the goal here is to argue, for researchers and educators alike, the need to think more widely about the nature of appropriate and ethical curriculum goals for STEM instruction.

Such an argument arises in no small part in reaction to numerous claims that STEM instruction will be a necessary component of an education for future economies, an argument that was also advanced for Science and Technology long before the STEM acronym existed (for example, NCEE, 1983; Williams, 2011; Zhao, 2019). Certainly, economic considerations must weigh heavily on public schooling, as a lowest common denominator of the public good schooling provides. However, I wish to take the progressive position of thinking about how we might make the schooling experience more educative. By this, I mean I wish for a vision of schooling that not only prepares its students for economic participation, but also participation in the civic and democratic processes of deciding what it is that populations ought to become in the future (see also Cowie & Mildenhall, this volume). We will need to reproduce many of the cultural achievements that are represented in part by our advances in understanding how the natural world behaves. However, my concern is that in our attention to reproduction, we do not pay enough attention to the production of new knowledge. The STEM disciplines will continue to be powerful ways of knowing (Maton & Moore, 2010; Young, 2008); possession of these knowledges will allow students to participate meaningfully in discussions about the future. Yet, there is concern here that these knowledges and ways of knowing carry with them particular value orientations that might limit the kinds of futures that we could conceive. If we understand particular conditions of our present time to be distasteful; if we can extrapolate current courses of action into the near future and predict disaster; if we can even suspect that our current ways of schooling may contribute to this state of affairs, should we not at least pause and consider what *other* goals we can ask schools to achieve, and how this might come about? In order that we might develop something approaching an action plan for STEM I begin by identifying some of the current excesses of Science and Technology.

10.3 What Are Some Problems with Science and Technology?

To avoid confusion, I shall use STEM to refer to the educational initiative that this book discusses, and Science and Technology (or S&T) as a shorthand to refer to the numerous *practices* that produce and make use of scientific, technological, engineering, and mathematical knowledge. I will consider reductionism, exosomatisation, and automation as three underpinning values of S&T, and will explain in detail how these values relate to some of the reasons why we may be pessimistic about the future. Before the discussion I note the essence of meaning of the labels of the three underpinning values I will consider.

Reductionism is the cognitive habit of S&T to seek simple explanations. Exosomatisation is the recognition that the human species is the only species that is so dependent on artificial devices and systems to ensure its continued survival. Finally, I use automation to refer to the technological tendency to erode the value of meaningful work in the pursuit of economic goals that benefit only a minority, something which further minimises our collective imagination of the possible. I will discuss these three values in some detail in subsections below. Before doing so, I explain why an exploration of these values is important.

While none are comprehensive nor completely representative, these three are some of the values that drive S&T and its deployment in many contemporary societies. What will be missing in the following discussions are any considerations about the nature of mathematics, in no small part due to my lack of familiarity with the matter. My ambition here is similar to earlier attempts to rethink science instruction by considering the nature of science (NOS). For instance, as the academic community became more aware of shifts in epistemic strategies in science, certain goals for science education became more, or less, important (Collins & Evans, 2017; DeBoer, 2013). Because STEM instruction now seeks to consider the holistic unity of phenomena and the deployment of these in societies, I contend that it is now important to consider the nature of technology and engineering, as it were, so that we might teach STEM appropriately.

A very significant difference between science and mathematics on the one hand, and technology and engineering on the other, is in the manner in which technology and engineering is more tightly integrated with societal desires and values. Certainly, we understand the pursuit of science to be linked to societal desires to understand; however, for the most part, scientists often do their work at a certain degree of remove from the public at large. Technologists and Engineers, however, have as their central project the design and creation of artefacts and systems that attend to human interests and give solutions to human problems. As a subject for school, STEM should not serve as merely an interesting vehicle of 'engaging activities and devices' to deliver the existing science and mathematics curriculum. Instead, the opportunity here is for students to learn about the technological world around them, in such a critical manner as to be empowered to make changes for a future which they will inherit.

In this regard, the nature of problem solving for societies requires some examination. While Science and Mathematics (S/M) are interested in problems of the natural world,[1] Technology and Engineering (T/E) can be said to be the science of the artificial (Simon, 1968/1996). Unlike the natural world, the world of artifice has different ways of defining its subjects of interest. While our interpretations of what constitutes S/M problems do change with time, they are definitely more durable than problems in T/E. Further, accepted solutions in S/M tend to resolve problems for a considerably longer period of time, and tend to have a context independent character. For example, electromagnetic laws discovered on earth also appear to function as well at the boundary of our solar system and beyond, as the space probe Voyager has amply demonstrated. Problems in T/E are more highly contextual, and commonly referred to as design problems. Design problems are characterised by their 'wicked' nature (Buchanan, 1992; Coyne, 2005; Rittel, 1972). Despite its connotation, wicked problems are so called not because they are evil, but because they resist solution; intriguingly, among other features, wicked problems: (i) cannot be solved for all time; (ii) have no definitive formulation; (iii) have no true-false solution, only better-worse; and (iv) can be considered symptoms of other problems. For instance, the characteristics of wicked problems were first noticed with social planning. Thinking about the ideal means of distributing a social good such as education will necessarily lead us directly into the unsatisfying domain of wicked problem solving. The design of things shares this 'wicked' nature, as we can see most recently with the numerous repeated iterations of technological gadgets.

If the nature of problems in T/E are wicked, and our value orientations lead us to interpreting problems in particular ways and thereafter to its solution, we can see that becoming aware of these values is the key to a vision of STEM education for

[1] There will be some controversy over the invented or discovered nature of mathematics I am certain (e.g., see Mansfield & Gunstone, this volume). For now, I will avoid it by artificially distinguishing between pure and applied mathematics, and considering only applied mathematics, as a language to communicate patterns in the sciences.

enlarged possibilities. If, for instance, we value capitalist orientations, we may be inclined to privilege economic growth no matter the cost, and so frame conversations about, for example, climate change in overly narrow terms. To teach STEM for humanistic goals requires us, in the first place, to consider how our current orientations may be detrimental to long-term human wellbeing, so that we can think about how else we should be orienting our ambitions of what we want our technologies to accomplish. STEM education, I am arguing, is as much an education in values as it is an education in its technical aspects, even for attempts to merely communicate the status quo. In other words, the status quo for STEM is not neutral, and instead communicates values, some of which may not be good for individuals and societies. The excesses of orientations that are the consequences of the three values discussed above are now considered.

10.3.1 Reductionism: Yes, But Where Are Its Limits?

Reductionism is an incredibly powerful orientation towards explanation. Instead of postulating the existence of multiple capricious deities responsible for various kinds of natural behaviour, it has been a grand achievement of post-Renaissance scientific thinking in establishing reductionism as the means of simplifying our analysis in order to allow our minds to grasp the core aspects of a phenomenon and its functions, its consequences, etc. An essentially unforeseen consequence of this form of thinking, however, has been that more complex systems are out of reach of these reductionistic forms of analysis. Education, for instance, is the interaction of multiple levels of phenomena, ranging from the innate species specific abilities, individual genetic endowment, psychological preference, classroom cultural patterns of behaviour, through to macro sociological descriptions of population level variation (see, e.g., Berliner, 2002; Luke, 2011; Phillips, 2014). Can we ever really be sure that, for instance, the findings of an educational intervention designed for children in one part of the world, and based on modifying one psychological variable, may be of service to children elsewhere? Certainly, we all have to find ways of coping, and trying anyway to do our best given the circumstances. However, the challenge for reductionism is what happens when we use it inappropriately.

Science and technology are really useful means of understanding, controlling, and predicting the natural world, but if we attempt to extend their methods and ways of framing and solving problems to where it may not apply, we can end up misunderstanding and oversimplifying phenomena. Even the relatively simple problem of what one should eat poses a challenge: reductionist thinking brings us the notion that there particular 'nutrients' that we ought to ingest, and that everything else is unimportant (Pollan, 2008). As a result, we can fall prey to well-meaning health advice advocating an increase (or reduction) of consumption of particular kinds of food, without paying enough attention to issues such as bioavailability, interactions with other elements of one's diet, the moderating effect of 'inactive ingredients', one's genetic heritage, and other factors. Distressingly, reductionism in eating advice also makes us susceptible to manipulative marketers who sell unhealthy food under the guise of containing particular nutrients that have been 'proven' to be healthy.

As science educators, it is certainly impossible to avoid reductionism, as it is a thinking tool of high utility. However, instead of allowing it to be communicated in a totally implicit manner as an unquestioned intellectual virtue, one that extends its validity across all other forms of inquiry, the minimum educators could do is to discuss its limitations where appropriate. Students should learn about epistemic habits in general: the different ways in which we come to know what we know, and how this and other habits may limit our access to truth. STEM, with an emphasis on real world problem solving, gives us clear opportunities. Demonstrations of the theoretical principles of Science often demand a very specific set of conditions. Even in mechanical constructions teachers often have to invoke 'friction' or 'non-uniformity' (or some other excuse) to explain deviations from 'standard behaviour', whereas in reality the principles only work in the ideal state which almost never occurs. Instead of blaming 'experimental error', teachers can move beyond approximations of ideality, and use realistic scenarios to acknowledge the complexity of the phenomena at hand. This is certainly not to say that teachers dismiss the achievements of scientists as being merely theoretical abstractions; it is to argue that, at appropriate times, teachers can illustrate the challenges required to move from messy realities to perfect abstractions and so enhance the achievements of scientists not diminish them. Such illustration may actually increase students' interest in science; in any case, it is a more accurate depiction of the work of science and scientists: natural laws do not lie 'out there' waiting to be discovered, but have to be teased out and carefully investigated and constructed by people.

10.3.2 Exosomatisation: What You Don't Know May Hurt You

To now shift the discussion from science towards technology, exosomatisation is the term used by the French philosopher Bernard Stiegler (Stiegler, 2018) to refer to exteriorisation of one's organs. By this, he means that we have become increasingly dependent for our existence on objects and systems that are not part of our bodies. For instance, we now have a range of technologies from warm clothing for individuals to collective power generation and heating systems that enable large population densities in regions not initially suitable for human habitation. It is not an overstatement to claim that we are all held hostage by the inventors and maintainers of these complex network of systems and processes that keep us alive. Consider the complex logistics networks required to make the warm clothing, or even the specific chain of methods required to transform raw material to end products: few of us actually know how the things we need are made, let alone know how to make the things we need. Even more so in our highly technologised societies, such dependency can be especially acute. Today we depend on motorised transportation, but few of us know what to do when things break down. The workings of the network infrastructure that sends our messages, ensuring that we will have food next week, and that the farmers get paid in return, similarly remains largely invisible.

The situation is much akin to what the iconic writer of science fiction Arthur C. Clarke observed: "any sufficiently developed technology is indistinguishable from magic" (Clarke, 1977). This situation is especially intolerable if we understand that the 'magicians' who work these systems can receive huge rewards in order to accomplish what can actually be fairly easily accomplished with the relevant knowledge typically kept obscure to maintain the economic disparities. Consider the conflicts over the right to repair a product of technology: from gigantic agricultural machines to the smallest electronic gadgets we rely on, manufacturers and consumer advocates have been fighting increasingly difficult battles over an individual's ability and right to repair and modify these products. While there are safety issues that can come from unauthorised modification (medical devices especially come to mind), it is hard to mount a defense against the argument that many manufacturers oppose increasing consumer rights simply because educated consumers can cut a manufacturing strategy of planned obsolescence and so continued profit.

STEM education, seen in this light, serves a liberating role as a means to metaphorically 'defrock the magician'. Indeed, such an approach already drives what has been commonly referred to the 'maker movement' (Blikstein, 2013; Dougherty, 2012; Martin, 2015), or the older 'DIY movement', associated with publications such as *Make* magazine, or, from an earlier time, the *Whole Earth Catalog* (Kirk, 2001). To accomplish this purpose of attending to the liberative goals of education, STEM classrooms can attend to what has been termed as 'epistemological dilution' (Papert & Harel, 1991). Seymour Papert, an early pioneer of the use of computers as tools for learning, was dismayed when he found numerous cases of teachers who used his software to teach students the very basic aspects of the use of the software, and nothing else, despite the richness of the programming environment. Today, we see examples of the same problem: schools introducing students to 3D printing via the activity of printing personalised keychains, and not going any further (Blikstein & Worsley, 2016). With STEM, the numerous possible forms of technology that we have make it very tempting to only use them in such very shallow ways. Instead of this shallowness we might want to consider a form of biological sciences approach, of treating these devices as 'laboratory specimens' for 'dissection', as exemplars of different 'species', to open up and study the commonalities and differences, but with a significant departure from the biological sciences approach, as I explain below.

While the language may perhaps sound slightly scary, teachers of technology can adopt what has been termed a hacker approach. Here, the term 'hack' is used in its 1984 definition by Stephen Levy (Levy, 2001): rather than illegal access to computer systems and theft of data, hacking in Levy's earlier meaning referred to the practices of technologists in the 1960s and 1970s, whose ethic of openness of access and freedom of information still live on today in software projects such as those that run on many contemporary information systems.[2] In addition to the ethic of openness and freedom of information, these hackers also had particular habits and practices which facilitated their investigations. For instance, as Simon surmised in the 1960s, designed objects exist at the boundary of the external desired behaviours, and the internal mechanisms which create these behaviours (Simon, 1968/1996). Think about a clock, for instance. Most of us merely make use of its external behaviour to tell time, and do not care if springs, or piezoelectric crystals, or network synchronised time drives the display. For hackers (in Levy's sense), this state is intolerable, because the workings have been made obscure, and because such a system prevents others from using the clock for other purposes (such as timed alarms). Thus, hackers often learn through a kind of 'reverse engineering', and, metaphorically or literally, take apart systems to uncover the inner working. To be sure, using these same thinking habits and practices on security systems *is* the first step to illegal trespass and nefarious behaviour. However, advising on ethical use of such skill can be managed. What would be considerably worse would be a circumstance where populations are deliberately prevented from inquiring into the very systems that sustain lives and livelihoods.

10.3.3 Automation: Artefacts Have Politics

When we think of automation, we often think about, for instance, production lines of robotic welders putting together parts for vehicles. Here, I expand the idea of automation to also include more contemporary iterations such as aircraft autopilot systems, intelligent decision assistance systems (for example those assisting physicians), and work scheduling systems such as those that run the 'gig economy'. These systems have certainly changed the way work is done and things are made, and liberated many workers from drudgery and offered many more people access to well made, inexpensive goods and services that were previously only available to a select few. Yet, there are rational grounds for questioning if some of these changes have actually been improvements to society as a whole, and where some shortcomings may be.

Nicholas Carr, in his insightful book on automation (Carr, 2014), argues that these automation systems have essentially deskilled human beings. Aircraft pilots are now put in an intolerable position where they have to watch over a machine which behaves predictably most of the time, and to take over when the machine stops doing so. Human attention does not function well in tasks like these, and so when autopilot systems fail, the human pilots may be unable to respond appropriately, with catastrophic consequences. Physicians working with electronic medical

[2] One significant example is the Android smartphone operating system. The MacOS kernel is also open source. Linux runs many servers around the world. Many 'smart' televisions today run some version of an open source operating system. Arduino is an open source hardware design—anybody can take the design and make as many copies of it as they like. And so on.

records and decision support systems may have reduced error rates in drug prescription and interactions. However, many have also been reduced to being just a human user interface connecting patient with computer system. Because physicians then do not have the time to truly interact with patients as human beings, diagnostic errors can increase. The gig economy brings us one step closer to a dystopian future where two classes of work remain: one which tells machines what to do, and the far larger group of people for whom machines tell them what to do.

This is not at all necessarily an argument for Luddism and the notion that there is something special about hand-made, artisanal craftwork, although there is an element of that. The more troubling issue here is with the relationship between technology and society, in that, in our quest for certain ideals of efficiency and profit, we can imagine and deploy systems which dehumanise and change our relationship with work and our interactions with people and things. From what was previously a holistic process in which skill, disciplined intuition, and human judgment were essential qualities to do work, we now have stripped work to become a programmed series of steps. Surely, this should be seen as the logical extension of both reductionism and exosomatisation, in that we not only have exported bodily processes (i.e., making things) to devices and systems, we have also reduced cognitive processes and become reliant on computers to think with.

There are two significant ways in which such a state of affairs is undesirable. The first is that the use of machines (and things in general) obscures the political intentions of its architect. When an 'impartial' system pronounces its judgment, we tend to accept its results as inherently more fair than when fickle/biased/possibly prejudiced humans do so.[3] Yet, as Winner (1980) and a host of Science-Technology-Society studies have shown (Wyatt, 2008), "artefacts have politics", as when nuclear power plants attract protests, or when bridges are deliberately made unnecessarily low in order to block the economically disadvantaged from access to particular neighbourhoods. In an instance of the latter example, the racial and economically marginalised who could not afford access to cars and so had to ride buses, were therefore denied access to Long Beach[4] because bridges were designed and built too low to permit passage of buses through key access roads. More recently, researchers have also pointed out how search engines and other 'big data' approaches to aiding decision making can be biased and discriminatory (Noble, 2018; O'Neil, 2017). Seen in this manner, it is not quite so much that we have to worry about dystopian futures where humans are being programmed by machines, but that we fail to question the underlying architectural assumptions and biases of those machines. If such objects have politics, they are the politics of the system architect and the designer. We should learn to interrogate such things for political intent, and to reject assumptions that there is a form of technical inevitability that explains why things have to be the way they are.

---

[3] Even more so if the machine produces numerical output.

[4] Long Beach, in New York, is the site of a state park, and a major site for recreation. Jones Beach, a widely acclaimed public park, was denied access to public bus riding people by over 200 low hanging overpasses. These tended to be low-income and ethnic minority.

The second reason why cognitive automation is undesirable is in the manner in which meaningful work is being gradually eliminated. When human agency is deleted, in big and small ways, work stops being meaningful for the worker and the human is now merely a step in a process that has yet to be automated. In the absence of purposeful employment, we really have to think hard about the consequent nature of our human existence: can we continue to serve our own purposes and goals, or are these to now be deleted in the service of other more powerful humans?

As I have argued above, the opportunity for STEM education is for a critical examination of the purposes to which we might put technologies. This requires a form of deconstruction that takes apart not only the technical functioning of the devices and systems around us, but also an analysis of the *purposes* that have been 'baked into' the technologies around us. In a similar manner, once we understand that these intentions are but one possibility for the use of a particular technology, we can then begin to see other possibilities for the deployment of the technology. In this manner, the approach I advocate here has some similarities with the critical literacy approach in other areas of schooling in which students are taught to analyse texts for bias, and then to rewrite the texts in different ways.

10.4 What Are Humanistic Goals for Education?

Above, I have summarised the ways in which S&T have contributed to the dehumanising aspects of our experience of contemporary living, in which the complexity and holism of life is simultaneously reduced and made obscure, and where people are trapped within webs of power in which their agentic autonomy is eroded, or even lost altogether. While issues of culpability are not the purpose of this chapter, it nonetheless is the case that educators can recognise that a range of possible outcomes exists for our students should we choose to pursue STEM. We can proceed with the status quo, and perhaps thus further intensify current problems, or we can modify STEM instruction to accommodate these problems in some way, with the knowledge that we might run the risk of creating further problems as we do so. This risk emerges because of the wicked nature of educational problems, but I argue that it is a risk certainly worth taking because I (and others) subjectively value the risk of status quo to be more damaging (see also, Roth & Désautels, 2002; Sadler, 2011; Zhao, 2019).

Here, I lay out an argument for a humanistic vision of education, broadly based on the notion that education should strive to give to students the greatest latitude for action in the future. While the humanist vision may be hard to clearly define, as many traditions exist, a useful beginning to developing such clarity here is to note what the humanist vision is not.

To begin, I do not believe that an ostensibly 'neutral' (especially STEM) education, based solely on communicating facts, is possible, nor is it desirable. In teaching STEM in a conventional curriculum, we already suffer from a "cult of relevance" (Conroy, 2020) in which decisions have been made concerning which facts are desirable. Frequently, this selection has been made based on concerns that curriculum needs to be relevant to students' futures. Such concerns for relevance are limited, ultimately to the experience of the planners, and importantly are ultimately limiting. Also, as mentioned above, economic development goals taken as a particular form of relevance are not sufficient, despite the overwhelming rhetoric argued in favour of this particular conception of relevance. The argument is commonly made that, for the benefit of the society as a whole, some form of schooling is needed to provide students with access to better paying jobs and to lift its poorest segments from poverty. At the other extreme, STEM is considered to be important for individuals and their future careers. While the latter may be the case, it is important to also consider the dignity of the individuals who will work these jobs, and the meaningfulness of the work that will be done. How well can we support S&T related jobs where humans are reduced to being human interfaces for bio-technological systems? How much can we support wealthy transnational organisations whose purpose is nothing less than the exploitation of mineral resources on one's land? In more contemporary times, the equivalent plunder is in their exploitation of human labour (Patel & Moore, 2017), their collection of vast amounts of personal data for profit and manipulation of 'consumers' (Zuboff, 2019), or their evasion of taxes. Certainly, there is potential for any small local firm to accomplish meaningful outcomes for its local community, but in order for that to happen, we need room to think differently from the hegemonic discourses surrounding the current positions about ideal deployment of S&T.

Educators should go beyond relevance and a concern for narrow, specific goals. Schooling has become accustomed to the logic of specific goals, especially those that may be standardised, tested, and compared (including internationally) in the industrial logic of benchmarking, such that its other functions are forgotten (Luke, 2011). This is a clear symptom of the dominance of a reductionist view of what *education* is good for: it becomes reduced to schooling/training, and then recently to a vision of ('self directed') 'learning'. Such a view privileges primarily the cognitive dimension and, conveniently, can be carried out largely by interaction with machines (e.g., Strauss, 2018). In this reduced vision of education, school becomes a place for downloading knowledge, and teachers merely the human interface for such machines (Williamson, 2016).

In a very broad conceptualisation, the philosopher Gert Biesta (2016) suggests three possible and diverse goals for education. As with the examples above, education can *qualify* people to work in jobs that they could not previously do. Education serves a *socialisation* function, in that one can obtain the skills to become part of established orders of being: what comes clearly to mind are overt nation building rituals such as flag raising and pledge ceremonies. More covertly, one can negotiate membership into communities such as the ethnic, gender, social, political, or economic elite by informal interactions within schools with large proportions of such individuals. However, most important for Biesta is the role schools can play in the *subjectification* of individuals. For Biesta, subjectification is the ability of individuals to stand apart from established orders of being, to imagine and achieve states of reality that do not currently exist.

It is this latter quality of subjectification that I believe schools can and should do better at. Here, I depart somewhat from a recurrent argument in science education that the numerous impending and actual catastrophes (for example climate change, global pandemics, loss of faith in science in general) demand our particular response in the classroom (e.g., Morin et al., 2017; Pedretti & Nazir, 2011; Sadler, 2011). While I agree we certainly need students to understand the issues and (especially) the scientific concepts under contention, it is a fine line between teaching these concepts, and recruiting our students to fight for our causes. Schools have to communicate both conservative and progressive positions, as we simultaneously need to not 'rediscover the wheel', and yet still have hopes of a different future. As Conroy (2020) asserts, such demands place "conflicting and contradictory expectations which result in [children] being ill-equipped to deal with the world as it appears in and to them" (p. 34). The school should instead be a site where we insulate children from the political conflict between adults, while helping to prepare them for participation. While Conroy recognises that opponents to this form of insulation will assert that there is no longer such a thing as a private space, he counters that such arguments for exposure are predicated on the false notion that exposure builds resilience:

> On the contrary, the creation of resilience may well require certain protections. Nowhere is this more evident than in the lives of those very vulnerable children who Prime Ministers are quick to identify as socio-political and economic problems [...] If such children are vulnerable because of their exposure to a range of behaviours including violence, sexual abuse, and drug and alcohol abuse, then might not the normative argument suggest that children should not be exposed to such things if they are to cultivate resilience. After all, we are generally not inclined to propose that we expose ourselves to a sexually transmitted infection so that we can build up immunity. Nor are we likely to suggest that children do a little experimenting with cocaine in the classroom. Exposure, even exposure with suitably and ideally presented discussion and reflection of the kind beloved of liberal educators, is not self-evidently the path to sustained resilience. (Conroy, 2020, p. 38)

Conroy also makes use of Hannah Arendt's notion of natality (Arendt, 1961): that every generation is born anew, with intentions and the full possibilities for action that even individuals in themselves are unable to foresee, let alone when acting in concert with others. Seen in this light, the purpose of school should be as a nursery is for plants: a warm and supportive environment that protects its seedlings from harsh environmental conditions outside. My humanistic vision of education is in agreement with Arendt, Biesta, and Conroy. It is a vision that schools should serve as sites where our children may find and create *their* own visions of the future, learning at the same time how to deal with the politics of this venture, while protected from the quarrels and contestations of the old world until such time that they believe they are ready to participate in the *polis*. To reiterate, I do not deny the importance of economic participation; acquiring qualifications and becoming socialised for particular kinds of careers is important. However, if this is all that school prepares students for, we will truly be sorry with the outcome.

10.5 What Is the Humanist Opportunity in STEM Education?

STEM, as disruption *du jour* to the normal functioning of schooling, offers us opportunities to rethink both our intended curriculum, and the pedagogical implementation of it. I want to suggest that the theoretical device of the discursive gap, previously proposed to consider the gap between theories of sociology and their phenomenological referents, can be particularly helpful for us here. Specifically, the role played by the term 'innovativeness' can be of service. STEM, as a curriculum goal for schools, has often been associated with innovativeness and economic development. Conventionally, this rhetoric of innovativeness is surrounded by simplistic notions such as the assertion that pure innovativeness explains how technology companies of Silicon Valley attained their globally dominant positions. Far from merely lifting themselves up through their bootstraps, this 'Californian ideology' (Barbrook & Cameron, 1996) overlooks many other contingencies and contradictions that make such levels of success outside of California questionable. For instance, the utopian vision of ostensibly meritocratic venture capitalist funding of ideas is reliant "upon a wilful blindness toward the other—much less positive—features of life on the West Coast: racism, poverty, and environmental degradation" (p. 45).

Nonetheless, innovativeness can still be reimagined and reclaimed by educators who wish for humanistic goals. In desiring novelty, economic innovativeness shares the humanist ambition to bring forth to the world what has not existed. Where teachers may assert themselves could be via the small, yet highly significant, role that can be played in addressing the ethical implications of the inventions that students will propose in STEM activities. Given the numerous learning goals that school teachers conventionally have to communicate in a school year, it is not surprising that STEM is seen, at best, as merely a vehicle for student engagement, if not an additional strain on their already overloaded schedule. Doing STEM in the way I suggest via this humanist perspective will only require teachers to simply slow down, and, as Arendt (1958/1998) would have it, to "think what we are doing". Here, I want to suggest that we collectively resist the temptation to 'operationalise' what such a pedagogy might look like, and instead trust *teachers as professionals* to have the appropriate judgment of what ought to be done in the classroom. If we accept that education is a complex activity, any attempt to develop a 'science' of education will inevitably reduce its richness. Subsequently despite all the best intentions, technological application of such a science will nonetheless run the risk of 'automating' the classroom.

We should, in the first place, consider what it is that we are doing with education. We are not, as the S&T metaphor may have it, transmitting knowledge just as we might be installing software updates on our computers. We are, instead, deliberately interfering with the creation of (forever) unfinished individuals, with all the attendant risk involved in the process. I use this term with the express understanding of its negative connotation simply as a reminder that schooling need not be *educative*.

As Biesta (2016) suggests, the risk is not that teachers or students are not good enough or that they do not try hard enough, but rather that if we are true to an education that is worthy of its name, we should be prepared for humans to exercise their right to refuse (see also, Labaree, 2004). Human interactions being what they are, there can be no guarantees that the best intentions will lead to good outcomes.

The role of the teacher in such an interpretation of education, including STEM education, is not as a production line assembler of 'learning outcomes', an end point in a technical-rational system optimised for a limited set of outcomes. Metaphorically, teachers need to be an artisan, competent in the ways in which their diverse palette of 'materials' may interact with the processes which they may subject it to.

Phronesis, an ancient Greek term referring to wisdom in practical action, is perceived in the quality of the teacher who is able to do the appropriate when faced with the complexity of education. Teachers will need to choose, with phronetic wisdom (Biesta, 2016; Flyvbjerg et al., 2012), the appropriate moment-by-moment pedagogical responses to their students' agentic actions, so as to maintain and create a wide range of possible outcomes for them. In other words, just as we now demand our STEM students to become creative and innovative, we have to demand the same of our teachers, and fully embrace the risk that this entails.

Teachers create the future, but this is not to be accomplished by an industrial process of carrying out a stepwise series of procedures that purport to guarantee particular outcomes. Rather, it is more like a *bricoleur*, starting with a general notion (and not a detailed plan) of a piece, and finding materials and interacting with them in unforeseeable ways depending on what the materials suggest the outcome to be. If we are to embrace innovativeness, we need to prepare for unexpected outcomes: herein lies the promise and opportunity for schools with STEM. As educators and researchers, we need to work with policymakers when they claim they desire innovativeness as an educational outcome. Surely we cannot fail to see the irony of the situation when we are asked to produce the equivalent of standardised creative outcomes. As STEM educators, the challenge and opportunity lies in correctly apprehending the epistemic practices and values of STEM, and communicating their limits accurately. These are already well known to the science (S) education community. However, now with innovativeness as a goal, we also need to appreciate the sociopolitical aspects of S&T, along with a healthy dose of ethical thought to its deployment. These aspects, I argue, are not as well developed, and (especially) science and mathematics educators may not be familiar with thinking about these aspects of their disciplines. In other words, with STEM conceptualised as curriculum and pedagogical disruption, there is now an opportunity to develop in teachers and students alike a different understanding of the purposes and goals for S&T. With innovativeness as a goal, we now have the license to move beyond sociopolitical reproduction, and to come closer to open ended humanistic goals for education.

Such an approach is likely to attract controversy for the potential unevenness of its outcomes and consequent implications for social justice. It is indeed the case that the industrial revolution and its methods of S&T have brought levels of luxury formerly accessible only to the elite to a substantially wider range of people. However, as I have shown above, these gains have not been without cost. Just as we can appreciate that the planet may not bear all the now multi-billions of humans who are collectively and continually diminishing its resources, we should consider if we really ought to desire everyone thinking the same way about S&T. To demand teachers and students think differently about the nature and goals of school can be hard, particularly given the historical conditions and the ways in which we have become entangled in the status quo. Yet we desperately need something other than the status quo of business as usual. With sincerity, I hope we will not be too late.

References

Arendt, H. (1961). *Between past and future: Six exercises in political thought*. Viking.

Arendt, H. (1998). *The human condition*. University of Chicago. (Original work published 1958).

Barbrook, R., & Cameron, A. (1996). The Californian ideology. *Science as Culture, 6*(1), 44–72. https://doi.org/10.1080/09505439609526455

Berliner, D. C. (2002). Comment: Educational research: The hardest science of all. *Educational Researcher, 31*(8), 18–20. https://doi.org/10.3102/0013189X031008018

Bevan, B., Gutwill, J. P., Petrich, M., & Wilkinson, K. (2015). Learning through STEM-rich tinkering: Findings from a jointly negotiated research project taken up in practice. *Science Education, 99*(1), 98–120. https://doi.org/10.1002/sce.21151

Biesta, G. (2016). *The beautiful risk of education*. Routledge.

Blikstein, P. (2013). Digital fabrication and 'making' in education: The democratization of invention. In J. Walter-Herrmann & C. Büching (Eds.), *FabLabs: Of machines, makers, and inventors* (pp. 203–221). Transcript.

Blikstein, P., & Worsley, M. (2016). Children are not hackers: Building a culture of powerful ideas, deep learning, and equity in the maker movement. In K. Peppler, E. Halverson, & Y. Kafai (Eds.), *Makeology: Makerspaces as learning environments* (pp. 64–80). Routledge.

Buchanan, R. (1992). Wicked problems in design thinking. *Design Issues, 8*(2), 5–21.

Capraro, R. M., Capraro, M. M., & Morgan, J. R. (Eds.). (2013). *STEM project-based learning: An integrated Science, Technology, Engineering, and Mathematics (STEM) approach*. Sense.

Carr, N. (2014). *The glass cage: How our computers are changing us*. W. W. Norton & Company.

Clarke, A. C. (1977). *Profiles of the future: An inquiry into the limits of the possible*. Harper & Row.

Clarke, D. (2014, July). Disciplinary inclusivity in educational research design: Permeability and affordances in STEM education. Keynote address to the international conference *STEM education and our planet: Making connections across contexts*. University of British Columbia. Retrieved from https://stem2014.sites.olt.ubc.ca/files/2014/07/Permeability-and-Affordances-in-STEM.pdf

Collins, H., & Evans, R. (2017). *Why democracies need science*. Wiley.

Conroy, J. (2020). Caught in the middle: Arendt, childhood and responsibility. *Journal of Philosophy of Education, 54*(1), 23–42. https://doi.org/10.1111/1467-9752.12367

Coyne, R. (2005). Wicked problems revisited. *Design Studies, 26*(1), 5–17. https://doi.org/10.1016/j.destud.2004.06.005

DeBoer, G. E. (2013). Science for all: Historical perspectives on policy for science education reform. In J. A. Bianchini, V. L. Akerson, A. C. Barton, O. Lee, & A. J. Rodriguez (Eds.), *Moving the equity agenda forward: Equity research, practice, policy in science education* (pp. 5–20). Springer.

Dougherty, D. (2012). The maker movement. *Innovations, 7*(3), 11–14.

Flyvbjerg, B., Landman, T., & Schram, S. (Eds.). (2012). *Real social science: Applied phronesis*. Cambridge University.

Kirk, A. (2001). Appropriating technology: The whole earth catalog and counterculture environmental politics. *Environmental History, 6*(3), 374–394. https://doi.org/10.2307/3985660
Labaree, D. F. (2004). *The trouble with ed schools*. Yale University.
Lehrer, R. (2016). Perspectives on elementary STEM education. In *Putting STEM education under the microscope*. Deakin University. Retrieved from https://blogs.deakin.edu.au/steme/wp-content/uploads/sites/39/2017/04/Lehrer-Deakin_STEM_2_reduced.pdf
Levy, S. (2001). *Hackers: Heroes of the computer revolution* (Vol. 4). Penguin.
Luke, A. (2011). Generalizing across borders: Policy and the limits of educational science. *Educational Researcher, 40*(8), 367–377. https://doi.org/10.3102/0013189X11424314
Martin, L. (2015). The promise of the maker movement for education. *Journal of Pre-College Engineering Education, 5*(1), 30–39. https://doi.org/10.7771/2157-9288.1099
Martín-Páez, T., Aguilera, D., Perales-Palacios, F. J., & Vílchez-González, J. M. (2019). What are we talking about when we talk about STEM education? A review of literature. *Science Education, 103*(4), 799–822. https://doi.org/10.1002/sce.21522
Maton, K., & Moore, R. (Eds.). (2010). *Social realism, knowledge and the sociology of education: Coalitions of the mind*. Continuum.
Moore, R., & Muller, J. (2002). The growth of knowledge and the discursive gap. *British Journal of Sociology of Education, 23*(4), 627–637. https://doi.org/10.1080/0142569022000038477
Morin, O., Simonneaux, L., & Tytler, R. (2017). Engaging with socially acute questions: Development and validation of an interactional reasoning framework. *Journal of Research in Science Teaching, 54*(7), 825–851. https://doi.org/10.1002/tea.21386
National Commission on Excellence in Education (NCEE). (1983). *A nation at risk: The imperative for educational advance*. U.S. Department of Education.
Noble, S. (2018). *Algorithms of oppression: How search engines reinforce racism* (1st ed.). NYU.
O'Neil, C. (2017). *Weapons of math destruction: How big data increases inequality and threatens democracy*. Penguin.
Papert, S., & Harel, I. (1991). Situating constructionism. In I. Harel & S. Papert (Eds.), *Constructionism* (pp. 1–11). Ablex.
Patel, R., & Moore, J. W. (2017). *A history of the world in seven cheap things: A guide to capitalism, nature, and the future of the planet*. University of California.
Pedretti, E., & Nazir, J. (2011). Currents in STSE education: Mapping a complex field, 40 years on. *Science Education, 95*(4), 601–626. https://doi.org/10.1002/sce.20435
Phillips, D. C. (2014). Research in the hard sciences, and in very hard 'softer' domains. *Educational Researcher, 43*(1), 9–11. https://doi.org/10.3102/0013189X13520293
Pollan, M. (2008). *In defense of food: An eater's manifesto*. Penguin.
Rittel, H. (1972). On the planning crisis: Systems analysis of the 'first and second generations'. *Bedriftsøkonomen, 8*(Norway), 390–396.
Roth, W. M., & Désautels, J. (2002). *Science education as/for sociopolitical action*. Peter Lang.
Sadler, T. D. (Ed.). (2011). *Socio-scientific issues in the classroom: Teaching, learning and research*. Springer.
Simon, H. A. (1996). *The sciences of the artificial*. MIT. (Original work published 1968.)
Stiegler, B. (2018). *The Neganthropocene* (Translated and edited, D. Ross). Open Humanities.
Strauss, V. (2018, November 17). Students protest Zuckerberg-backed digital learning program and ask him: 'What gives you this right?' *The Washington Post*. Retrieved from https://www.washingtonpost.com/education/2018/11/17/students-protest-zuckerberg-backed-digital-learning-program-ask-him-what-gives-you-this-right/
Tytler, R., Williams, G., Hobbs, L., & Anderson, J. (2019). Challenges and opportunities for a STEM interdisciplinary agenda. In B. Doig, J. Williams, D. Swanson, R. Borromeo Ferri, & P. Drake (Eds.), *Interdisciplinary mathematics education: The state of the art and beyond* (pp. 51–81) Springer. https://doi.org/10.1007/978-3-030-11066-6_5.
Williams, P. J. (2011). STEM Education: Proceed with caution. *International Journal of Technology and Design Education, 16*(1).

Williamson, B. (2016). Silicon startup schools: Technocracy, algorithmic imaginaries and venture philanthropy in corporate education reform. *Critical Studies in Education, 59*(2), 218–236. https://doi.org/10.1080/17508487.2016.1186710

Winner, L. (1980). Do artifacts have politics? *Daedalus, 109*(1), 121–136.

Wyatt, S. (2008). Technological determinism is dead; long live technological determinism. In E. J. Hackett, O. Amsterdamska, M. Lynch, & J. Wajcman (Eds.), *The handbook of science and technology studies* (pp. 165–180). MIT.

Young, M. (2008). From constructivism to realism in the sociology of the curriculum. *Review of Research in Education, 32*(1), 1–28. https://doi.org/10.3102/0091732X07308969

Zhao, Y. (2019). The rise of the useless: The case for talent diversity. *Journal of Science Education and Technology, 28*(1), 62–68. https://doi.org/10.1007/s10956-018-9743-3

Zuboff, S. (2019). *The age of surveillance capitalism: The fight for a human future at the new frontier of power*. Public Affairs.

Chapter 11
Final Commentary: "Education in the 21st Century: STEM, Creativity and Critical Thinking"

Amanda Berry

Abstract This chapter offers a reflective commentary on the contribution of the book as a whole, taking up the idea supported by the book's authors that STEM education can, and needs to be, much more than an educational reform agenda to supply a future global workforce. Based on the work of the authors, a set of key themes is identified and discussed that can progress the development of STEM education, nationally and internationally. These themes include: embracing contemporary views of education, developing skills and capabilities such as critical thinking and creativity, engaging with societal issues of social justice and equity (including the role of empathy), and better understanding the nature of teacher expertise and its development.

Keywords STEM education · Teacher professional knowledge · Policy · Collaboration · GERM theory

11.1 Introduction

In this final chapter, I offer a reflective commentary on the contribution of the book as a whole, taking up the idea supported by the book's authors that STEM education can, and needs to be, much more than an educational reform agenda to supply a future global workforce. It was intentional in our conceptualisation of this book and in the workshop discussions that the chapter authors would offer a perspective of STEM education that includes a much broader and more inclusive vision of schooling and education than STEM for economic and employment purposes. This vision embraces the view that STEM education can, and should, contribute to a better global society and its citizenry, through preparing students who can effectively respond to multi-faceted economic, social and environmental challenges, such as

A. Berry (✉)
Faculty of Education, Monash University, Clayton, VIC, Australia
e-mail: amanda.berry@monash.edu

A. Berry et al. (eds.), *Education in the 21st Century*,
https://doi.org/10.1007/978-3-030-85300-6_11

those foregrounded by COVID-19 and climate change (Corrigan, 2020). In this way, we distinguish between notions of STEM as a workforce development initiative, and STEM education, as distinct and necessarily different terms.

Emerging from the book's chapters is a set of key themes that we hope will help to progress the development of STEM education, nationally and internationally. These themes include: embracing contemporary views of education, developing skills and capabilities such as critical thinking and creativity, engaging with societal issues of social justice and equity (including the role of empathy), and better understanding the nature of teacher expertise and its development. While individual chapters may address one or more of these themes, at the heart of this book, it is clear that none of these entities exist in isolation. They are deeply entwined. Understanding how they are entwined, and in what contexts, and how they are central to STEM education, is a consistent focus throughout the chapters.

11.2 STEM: A Phenomenon of Global Interest

In Chap. 1, we outlined how the notion of STEM has rapidly captured international interest and emerged as an educational phenomenon. The global focus on STEM has been driven primarily by external forces to serve economic and vocational goals, funded by governments and typically promoted by politicians and industry (Blackley & Howell, 2015). As a consequence of the involvement of different interest groups and their agendas for STEM, different meanings of STEM have emerged and been pursued, including STEM as a set of career fields, a collective noun for the separate disciplines of the sciences, mathematics, engineering and technology, and as an integrated approach to curriculum (often referred to as integrated STEM). These different meanings have subsequently led to varied STEM narratives that have, in turn, influenced the kinds of strategies and programmes that have been pursued in schools and other educational contexts.

For example, in Australia (my home country), concerns about a limited supply of future STEM workers has led to STEM becoming imposed on schools as an employment problem to solve. Added to this, falling interest and participation rates in senior science and mathematics and low levels of student achievement against international benchmarks, have led to a discourse of "STEM in crisis" (Marginson et al., 2013, p. 55). Proposed solutions typically include adding more science and mathematics into school programmes, a focus on student mastery of disciplinary content, and recruiting more teachers with strong disciplinary backgrounds in science and mathematics. However, more recently, the STEM crisis has been labeled as "a myth" (Ritchie, 2019) that was created to drive policy to promote STEM education that would address future workforce demands and increase achievement in cross-country comparisons resulting from high-stakes international testing (e.g., TIMSS and PISA), results acting as a proxy for quality education.

STEM as a global education policy is now further explored.

11.3 STEM: A Kind of "GERM"?

In Chap. 1, we referred to the dramatic and devastating effects of the novel corona virus (COVID-19) on humanity. As a rapidly spreading and hard to control infection, COVID-19 has vastly changed our ways of living in a very short period of time. An important factor contributing to the rapid spread of COVID-19 is globalisation: the increased, large-scale mobility of people and goods around the world. Similarly, globalisation has been an important factor influencing the spread of large-scale educational reforms, such as STEM. In education, these reforms are typically driven by a focus on improved educational performance through factors such as competition and standardisation. Sahlberg (2011) coined the metaphor-acronym 'GERM' — Global Education Reform Movement — to describe the pervasive spread of reforms through education policy that "infects education systems" worldwide. STEM education is an obvious example of a GERM. Promoted as an educational imperative that can improve national competitiveness, drive economic expansion and transform society, the STEM education agenda is highly influenced by forces from outside of education (e.g., government, industry) with significant impacts on schools and teachers.

Sahlberg (2011) described six features of GERMs that have aimed to improve education but that have problematic side effects:

1. Standardisation of, and in, education: This shifts the nature of teaching from an open-ended, non-linear process of mutual inquiry and exploration to a linear process with causal outcomes. Standardisation may also restrict creativity and innovation in teaching and learning.
2. Focus on literacy and numeracy: This sets up silos in these areas and reduces focus on other subjects.
3. Teaching for pre-determined results: This minimises risk-taking in teaching and learning and, therefore, reduces opportunities for creativity.
4. Market-oriented reform ideas: These distance teachers from the moral purpose of their profession.
5. Test-based accountability: This increases instances of teaching to the test.
6. Control: This diminishes autonomy of teachers and the degrees of freedom of schools. This can lead to teaching that aims to showcase good practices rather than genuinely helping students to learn.

Several of these GERM features are evident in efforts to drive STEM education. For example, focusing measures of success in terms of student achievement in individual STEM subjects, assessing STEM features such as critical and creative thinking as specific measurable entities, focusing on the products of STEM rather than its processes (e.g., in terms of students producing products for market), and collapsing the STEM disciplines into one entity, thus distorting views of their distinctiveness (e.g., the notion of failure as the same in each of the STEM disciplines, c.f. Mansfield & Gunstone, Chap. 9). If we are to avoid succumbing to the side effects associated with a GERM agenda, framing what STEM can, and should, look like in education requires engaging with fundamental questions about the purposes of contemporary schooling and what makes a *good education*, including the role of teachers in educational reform. Rather than essentialising STEM as a high-performance competition or training for industry, a "STEM education" needs to take into account broader society aspirations:

> Framing the primary purpose of schooling and STEM course work in terms of job preparation, economic growth and national security is problematic… [These] are, at best, only very loosely tied to the general state of schooling, and the need for a technical workforce does not provide a compelling impetus for most children to value STEM learning… a K-12 system focused on job preparation cannot keep up with the ever-shifting job market and would ill-prepare such individuals when such changes inevitably occur. … A STEM education, as opposed to a mere training, would draw from and dignify the humanities in a common effort to prepare individuals for engaged citizenship which includes judiciously assessing the pros and cons of STEM for improving personal and societal welfare. (Zeidler et al., 2016, p. 466)

In the sciences, there is a long history of the failure of major curriculum reform initiatives aimed at increasing the post-school scientific workforce. In fact, the best school science preparation for potential scientists in contemporary times has shown to be the same broad and socially contextualised form of school science education that has been argued across the last five decades as best for those *not* planning a scientific career (Smith & Gunstone, 2009).

The contributors to this book offer a very different vision of the purposes and processes of STEM education. This vision transcends a GERM like interpretation, and may even offer ways of vaccinating teachers and schools against a view of STEM as GERM. Through their individual chapters, the authors collectively highlight an important theme: that STEM education in formal and informal contexts has potential to offer an approach to working with, and preparing, young people to actively, thoughtfully and productively participate in all aspects of society through engaging their interests in, and commitment to, STEM. This view of STEM as enabling young people's informed, civic participation in Science, Technology, Engineering and Mathematics has important implications for teachers, students and schooling.

11.4 What the Chapters Show Us About STEM Education

The chapters of this book show us that STEM education can (and should) be defined in different ways according to need and context; that STEM education goes beyond the teaching of the individual disciplines within their silos; and that STEM education involves learning and using different ways of thinking and knowing, collaborating across and within different groups (within and outside of schools), and learning about and taking risks. Each of these features of STEM education is elaborated below, along with illustrative examples from each of the chapters.

11.4.1 STEM Education Should Be Defined According to Need and Context

One of the challenges of STEM education is in the label itself. What does 'STEM' actually mean in the context of education? Earlier I pointed out that STEM is broadly associated with the individual disciplines of its acronym, or as a set of career fields. In schools, STEM education has taken on a variety of meanings, including the teaching and learning of one or more of the four individual disciplines, or in an integrated way (in combinations of two, three or four of the disciplines); and/or a focus on particular skills and capabilities, such as critical thinking and creativity, or metacognitive skills, such as learning to monitor one's own learning. This multiplicity of meanings and applications of STEM in education is both problematic and useful. On the one hand it allows stakeholders, within their particular and unique contexts, to define STEM in ways that best suit their needs; on the other, it may lead to STEM education being interpreted as 'anything goes'. For example, some interpretations of STEM emphasise the importance of including lots of 'hands-on activities', which research has demonstrated may actually reduce opportunities for students' meaningful learning when these are the primary or sole focus (see, for example, Berry et al., 2001).

One issue taken up by contributors to this book is how the meaning of STEM education has been interpreted through the curriculum. The development of STEM education curriculum should plot a path to be taken, rather than a detailed syllabus specifying exactly what teachers need to do. For example, Buntting and Jones (Chap. 5) point out that in the New Zealand context, the national curriculum provides a broad set of learning aims and attributes to be developed, such as innovation, inquiry and curiosity; and key competencies, including creative and critical thinking and relating to others. Within this national curriculum, schools and teachers have the autonomy to develop locally relevant STEM curriculum and are encouraged to make natural connections across learning areas, values and key competencies.

Rennie (Chap. 7) expands on the important role of curriculum, recalling Ralph Tyler's (1949) notion that curriculum should help in realising the goals of education through asking, *What are the purposes of learning? How can learning experiences be organised and evaluated to realise these purposes?* Rennie distinguishes between two dimensions of curriculum: the balance between disciplinary and integrated knowledge (where 'knowledge' is used holistically to include accompanying skills and capabilities), and the balance between local and global types of knowledge (Rennie et al., 2012). Any curriculum should offer balance and connection across both of these dimensions. Elsewhere, Rennie (1990) has emphasised the need for students to "experience a meaning and a context for what they have the opportunity to learn." Otherwise, "they are unlikely to learn it" (p. 191). Importantly, both Chap. 5 and Chap. 7 highlight the more open notion of curriculum as a framework to be developed in context, rather than positioning curriculum as a heavily specified pathway in a syllabus.

Critical to realising a relevant and meaningful STEM curriculum is the opportunity for building a shared understanding of its purposes. This point underpins Chap. 2 by Kelly and Ellerton and Chap. 3 by Vincent-Lancrin. Kelly and Ellerton point out that the combination of the concepts of creativity and critical thinking provides a very potent avenue for integrated STEM educational practices. Establishing a shared understanding of each of these concepts and their mutual dependence is developed throughout the chapter: critical thinking permeates every aspect of creative practice, while creative practice catalyses the growth and complexity of critical thinking.

In Chap. 3, Vincent-Lancrin reports on an OECD project that seeks to foster and assess creativity and critical thinking through identifying some of the elements that could lead to a common understanding of these ideas, and how teaching and learning strategies aligned to the development of creativity and critical thinking can be applied in both domain-specific (science) and domain-general ways. Vincent-Lancrin describes two domain-general conceptual rubrics for assessing creativity and critical thinking, including a "comprehensive" rubric and "class-friendly" rubric. An important purpose for creating these rubrics is to assist teachers to apply their understandings of creativity and critical thinking *to their own teaching and learning contexts*. To further support teachers, Vincent-Lancrin outlines a set of "design criteria," including motivation, cognitive activation, and self-regulation, as well as opportunities for formative assessment that teachers can consider when designing science learning experiences that support their students to develop students' creativity and critical thinking capabilities. While the design criteria are important, successful teaching of creativity and critical thinking also relies on teachers' attitudes and abilities to create suitable learning environments where risks, failure and mistakes are a naturally accepted part of the learning process (see Chap. 5 by Buntting and Jones; issues of risk and failure in science and mathematics learning is taken up in detail in Chap. 9 by Mansfield and Gunstone).

The influence of context on how teachers define and put into practice creativity and critical thinking is further elaborated in Chap. 6, by Corrigan, Pannizon and Smith. In their study of developing creativity and critical thinking across early childhood, primary and secondary school networks, they found that context shaped the pedagogical approaches teachers utilised to develop these capabilities in their students. In other words, the age of the children made a difference. For example, in early childhood settings, when asked to explicitly develop creativity and critical thinking, teachers tended to focus on open-ended approaches that were both planned and incidental in nature. In primary schools, problem-based learning and thinking routines featured prominently, while in secondary schools, teachers sought to develop a shared understanding of creativity and critical thinking and considered how they might be applied in specific disciplines. The findings from Corrigan, Pannizon and Smith's study highlight the need for teachers to translate the concepts of creativity, critical thinking and the integrated nature of STEM into their practical realities. Another key insight is that, as the teachers reflected on their current ideas and practices and sought to incorporate opportunities to further develop students' thinking, either by drawing on existing approaches or developing new ones, they began to work more collaboratively, clarifying their individual and collective understanding of these ideas and how to implement them.

Clearly, across the different interpretations of STEM education described by the authors of this book, there is an emphasis on the need for teachers to be actively involved in decisions about what matters and why in their own contexts. This active decision making of teachers needs to include how to incorporate the learning of different skills and capabilities within a STEM education curriculum. Of course, having some support and guidance is helpful for teachers as they develop and implement curriculum change, but that is not the same as a prescriptive, one-size-fits-all template. In this respect, teachers and schools need to be trusted to make decisions about what is best for their students.

The different examples offered by the authors of this book also highlight the important notion of STEM as "a collection of evolving ideas rather than a specific approach or practice" (Siegel & Giamellaro, 2019, p. 757). STEM education cannot rely on any specific approach or practice, as each context in which it is enacted will be unique.

11.4.2 STEM Education Goes Beyond the Individual Teaching of Disciplines to Become "Ways of Thinking and Working"

The contributors to this book propose a view of STEM education that goes beyond the boundaries of its individual disciplines to embrace a more local, context-driven, interdisciplinary approach that draws on and connects particular ways of thinking and working. However, as Tan points out (Chap. 10), the traditions of curriculum and schooling may make such a goal difficult to realise. For example, long-standing and deeply-engrained views of what comprises each of the individual STEM disciplines, including what students need to learn and in what order, particularly in science and mathematics, have a powerful effect on shaping the structure of curriculum and schooling, and have shown remarkable resistance to efforts for change.

Going beyond the traditional structures and ways of working of individual STEM subjects also requires knowing how to meaningfully incorporate opportunities for developing creativity and critical thinking into STEM learning. As Kelly and Ellerton (Chap. 2) note, "It is important to view creativity and critical thinking in concert to fully understand their interrelatedness and the educational ecosystem they enable to maximise educational potentials in STEM education". Kelly and Ellerton also recognise the tensions associated with existing disciplinary traditions and raise the question of "how to operationalise creativity in educational practice against a backdrop of such traditional educational discourses".

In order to take an integrated approach to STEM education, Buntting and Jones (Chap. 5) highlight the need for teachers to be able to understand and navigate the different discourses of the STEM disciplines. Through the case study of one teacher,

James, their chapter explores the professional knowledge and skills that teachers needed to scaffold students' STEM learning, and the value of focused learning conversations with students to support their creativity and critical thinking. Buntting and Jones also remind us of the complexity of this task for teachers, who need to be able to recognise and navigate the different scientific, mathematical, technological and everyday discourses, and know which discourse to use when and where, if they are to support students' conceptual and skill development throughout a STEM sequence of learning. This could be a key aspect to be addressed in teachers' professional learning and development.

Corrigan, Panizzon and Smith (Chap. 6) similarly advocate the need for teachers and students to understand the different forms of expertise embedded within and across the STEM disciplines. Their study provided opportunities for teachers to investigate differences between the STEM disciplinary areas and build an appreciation of how working *across* the STEM disciplines can open up new possibilities for student learning. Corrigan et al. also identified that while there is a growing emphasis on developing students' ways of thinking, teachers tend to perceive these as 'add-ons' to the curriculum, and they are often applied in a tokenistic rather than integrated manner. Here, rubrics for creative and critical thinking, such as those discussed by Vincent-Lancrin (Chap. 3) may assist teachers to embed these ways of thinking more purposefully within the curriculum.

Seeing what the integration of the STEM disciplines can look like and how this can be operationalised in practice is a focus of Rennie's work (Chap. 7). Her chapter explores how effective integrated curricula with an out-of-school component encourages students to develop their STEM understanding and skills. Three important aspects of this approach are highlighted:

1. using real-world, authentic contexts that can meaningfully bring together disciplinary and interdisciplinary knowledge;
2. working outside of the regular classroom to show students a "bigger picture"; and
3. working on issues that are important to the local community, and/or matters relating to social values and diversity, provides students with opportunities to develop their senses of social and ecojustice.

Through such integrated learning opportunities, associated thinking skills and capabilities can be developed. Drawing on the OECD's dimensions of creativity and critical thinking – inquiring, imagining, doing, and reflecting – can be helpful to illustrate how guiding students to interact with local, place-based, or community issues can enhance their creativity and critical thinking, as well as their communication and collaboration skills.

The three aspects highlighted by Rennie above, are also evident in the chapter by Cowie and Midenhall (Chap. 4), who also show how operationalising the STEM disciplines through integrated approaches can play a role in developing students' sense of social justice. Through the presentation of three vignettes of classroom practice, Cowie and Mildenhall illustrate how approaches that incorporate the development of knowledge, empathy and action, can support students' capacity for critical and creative thinking and the inclination to take constructive action for wider societal 'good'.

Cheng and Leung (Chap. 8) extend the idea of socio-scientific aspects of STEM education into higher education through their example of an interdisciplinary course that aims to engage students in "a critical scrutiny of their thinking and of the information they come across," focusing on the specific issue of obesity. Cheng and Leung highlight how working across the STEM disciplines helps students to develop an appreciation for different ways of thinking, for example, technocratic ways of thinking (based on scientific rigour) versus emancipatory thinking (based on ethical and political scrutiny of an issue). Their approach provides an interesting example of what it means to operationalise student learning in more transdisciplinary ways, where the boundaries between different disciplines and ways of thinking are challenged and become intentionally blurred.

Collectively, the book's examples described above provide a range of possibilities for moving beyond STEM as 'siloed'. However, it is also important not to lose sight of the value of the individual disciplines themselves and their unique ways of working and thinking. Indeed, the contributors to this book are not suggesting blending the individual disciplines into what Lehrer (2016) calls an "epistemic stew," but instead, becoming cognisant of what each discipline contributes and how the disciplines can be meaningfully drawn upon and connected in the development of student learning.

11.4.3 STEM Education Involves Collaboration

The chapters of this book illustrate that collaboration is an important component of STEM education, and that collaboration looks different according to the learning purpose(s) and context. For example, collaboration can include teachers within a department or school, across schools or across school sectors (e.g., schools and universities), between students within or across grade levels or schools, and between schools and communities/industry in their local or broader contexts. However, popular interpretations of collaboration are often rather loosely defined as 'simply' working in groups. Collaboration needs to be purposefully planned, drawing on different kinds of social and cultural activities, such as learning to work with others, engaging in active discussion and shared decision making, and joint problem solving.

The OECD (2005) identifies collaboration as a multi-faceted key competency for the twenty-first century. It requires:

- the ability to relate well to others, including demonstrating empathy and effective management of one's own emotions;
- the ability to cooperate in terms of presenting ideas and listening to those of others, understanding the dynamics of debate, being able to follow an agenda, being

able to construct tactical or sustainable alliances, being able to negotiate and make decisions that allow for different shades of opinion; and
- the ability to manage and resolve conflicts where issues are analysed for different interests, such as power, equity, recognition of merit and division of work, the origins of conflict, the reasoning positions of different sides and recognition of different possible solutions, being able to identify areas of agreement and disagreement, to reframe a problem, and prioritise needs and goals including deciding what you are willing to give up and under what circumstances.

Chapter authors provide insights into different kinds of collaboration, why it matters, and what can be learned through collaborating. At the most fundamental level, Kelly and Ellerton (Chap. 2) highlight the social collaborative nature needed for STEM education in terms of the development of students' collaborative and communication capacity. Other chapters provide examples of how collaborative relationships link with particular learning purposes. For example, Rennie (Chap. 7) describes how a school and local industry collaboration supported students' learning about local environmental issues, where the collaboration was developed between different groupings of students, teachers, experts, parents and the general public. These varied types of collaboration helped students to develop multiple project outcomes as well as to negotiate decisions about which solution(s) would be the most appropriate to pursue. Cowie and Mildenhall (Chap. 4) highlight a collaborative relationship between students and a local community group that resulted in students developing a strong sense of community and that their ideas were listened to and mattered, while the members of the community group gained a renewed sense of purpose in helping others. Buntting and Jones (Chap. 5) alludes to the benefits of collaboration between teachers and researchers, as well the more obvious teacher/student and student/student collaborations. Corrigan, Pannizon and Smith (Chap. 6) highlight how collaboration across the early childhood, primary, secondary and tertiary sectors supported teacher and researcher learning about effective pedagogies. Specifically, sharing professional knowledge beyond individual teachers and sites helped to build collective capacity and shared understanding of ideas that not only benefited the teachers, but that also provided a model for how students might work together. In a similar vein, Cheng and Leung (Chap. 8) demonstrate how collaborations between university educators across different faculties can support student learning in a general university course based on a socio-scientific issue.

Overall, the role of collaboration highlights STEM education as a sociocultural activity that is simultaneously "active, contextual, co-constructed, and continually evolving" (Siegel & Giamellaro, 2019, p. 768). However, thinking about STEM in this way also presents challenges for its implementation, where traditional educational structures and processes present barriers in terms of (for example) time tabling, non-alignment of curriculum, and the need to specify and standardise predetermined learning outcomes rather than allowing these to evolve through a project. Seeing STEM education as a sociocultural activity that involves co-construction and that evolves over time according to need, interest and resources, requires system flexibility that the chapter authors show is possible – but that requires commitment and motivation of both teachers and educational leaders.

11.4.4 STEM Education Involves Risk

Biesta (2013) identified that

> ...real education always involves a risk...The risk is there because, as W. B. Yeats has put it, education is not about filling a bucket but about lighting a fire....The risk is there because students are not to be seen as objects to be moulded and disciplined, but as subjects of action and responsibility. (p. 1)

If 'real education' involves risk, it is tragic that education systems are typically set up for risk aversion. This is evident through the many control mechanisms built into the schooling system, for example, through a tightly regulated curriculum, and standardised assessment and reporting procedures that keep teachers and schools highly accountable for their performance. Under such conditions, the opportunity for "real education" in Biesta's terms, becomes very difficult. Viewed from a STEM perspective, a significant tension emerges between what is often promoted as central to STEM education, that is, engaging with risk and failure (e.g., through design and prototyping), and the performance-related, risk-averse nature of formal education. Many of the authors in this book draw attention to these issues as they discuss the role of failure and risk.

It is not only the general nature of schooling that makes engaging with risk and failure difficult, but as Kelly and Ellerton point out (Chap. 2), the discourses within the STEM disciplines themselves tend to emphasise knowledge as transferable "and corresponding assessment of learner retainment of this knowledge". Such traditional views of knowledge acquisition and reproduction make it difficult to operationalise creativity in STEM Education, which "requires a highly interactive, experiential culture with low risk-aversion" that can support "collaborative ideation and prototyping over time in any discipline context".

Mansfield and Gunstone (Chap. 9) delve into the ways in which knowledge is represented and developed in each of STEM disciplines, including the role of failure. They point out the serious "misalignment" between the ways in which the role of failure in each of the STEM disciplines "is represented to and perceived by those outside the discipline", and also in the ways that the individual STEM subjects are taught and learned in school. For example, the field of science is typically viewed by the general public in terms of facts and irrefutable truths, a view that is also reflected in traditional school science education through the predominance of 'recipe style' labs with pre-determined outcomes, and assessment tasks that value recall of facts. Such representations of science sit in sharp contrast with how the field of science actually progresses ("involving struggle, failure and/or serendipity") and an approach to STEM education advocated by the contributors of this book, i.e., STEM education that encourages uncertainty, doubt and risk. Interestingly, Mansfield and

Gunstone note that the concept of failure rarely appears in accounts of the development of mathematical knowledge and that school mathematics has tended to be associated with "fear of failure" (also known as "mathematics anxiety"). Thus, they advocate for including more opportunities for school students to "recognise the role and value of failure in both real-world STEM contexts and in their own learning of STEM") that may contribute to increased awareness of its meaning, value and prevalence in these fields. Their advocacy to view failure as a form of success offers both encouragement and a way forward for teachers and learners.

Several other chapters also point to the challenges for teachers in supporting student learning about failure, uncertainty and risk. While managing a risk-averse system is one significant challenge, another lies in teachers feeling sufficiently confident and prepared to teach about risk and failure to students. Kelly and Ellerton (Chap. 2) and Vincent-Lancrin (Chap. 3) link the notion of risk to teaching and learning:

> The successful teaching of creativity and critical thinking also hinges critically on teachers' attitude and in their ability to create learning environments where students feel safe to take risks in their thinking and expressions. This in turn presupposes a positive attitude towards mistakes and learner empowerment. (Kelly & Ellerton, Chap. 2)

The risk for teachers is real, but as Tan (Chap. 10) points out, if we require our STEM students to become creative and innovative, we need to support our teachers to be the same, and to be able to fully embrace the potential risk that this entails. Failure needs to be seen both as a real possibility, and as an opportunity for learning. If this is not the case, why would new ideas and approaches to be tried? The need to trust in teachers' professionalism is regularly being eroded in many educational settings through regulation and control mechanisms that limit opportunities for experimentation and risk – a feature of GERMs, as outlined earlier.

Risk-aversion in educational systems is a long-standing and pervasive challenge, even though risk and uncertainty has always been an essential part of science (see Rennie, 2020). Calls for including learning about the nature of knowledge as uncertain and as an integral part of school science education have been outlined by Fensham (2011), who detailed a Cynefin[1] framework to assist teaching complexity through certainty and uncertainty. This is but one example that teachers could adopt. STEM education can also highlight the important value and role of failure though more authentic experiences. When there is shared understanding of what failure may offer, if it is forward-focused and can be seen as an opportunity, students can gain a more authentic appreciation of the STEM disciplines.

---

[1] Cynefin is a Welsh word for multiple locations

11.5 Building an Alternate Interpretation of STEM Education

Based on the idea that the current 'crisis' discourse around STEM appears to narrow possibilities for learners and learning, it is my hope that this final chapter helps to bring together the messages of the book to illustrate how STEM education might be differently interpreted to broaden its possibilities and to benefit a broader range of learners. In the following section, I highlight some of the implications of STEM education for schools, teachers and learners.

11.5.1 STEM Education Opens Up Important Opportunities for Schools, Teachers and Learners

11.5.1.1 School Organisation

School organisation needs to be challenged as 'one size' does not 'fit all'. Schools need to find new and novel ways to operate that are responsive to the needs and opportunities available within their local communities. As Tan (Chap. 10) points out, "seriously integrated STEM will best come from the ways each individual school plays to its teaching and resources and environment strengths". Such new ways of working should also include empowering and trusting teachers as curriculum developers and decision-makers.

11.5.2 Empowerment of Teachers

The STEM education agenda can, and needs to be, utilised as a means to enable and empower teachers to act as agents of pedagogical change and transformation, and to enrich teachers' professional knowledge of practice. The ideas of this book may help to progress this agenda, through:

- the small, yet highly significant, role that can be played in addressing the ethical implications of the inventions that students will propose in STEM activities (see Tan, Chap. 10);
- the role teachers play in utilising local and global STEM contexts for building shared learning opportunities in creative and critical thinking (see Rennie, Chap. 7);
- the development of teacher expertise to facilitate targeted learning conversations to support students' conceptual and procedural learning outcomes (see Buntting & Jones, Chap. 5); and
- the development of new knowledge of STEM as opposed to the epistemological differences that exist across S, T, E and M (see Mansfield & Gunstone, Chap. 9).

The vision of STEM education and the kinds of school and teacher change inherent in this agenda cannot occur without significant shifts, including shifts in conceptions of what to teach, conceptions of identity as teacher, conceptions of learning and learners, and conceptions of the nature of teacher professional expertise. In turn, these shifts need to mutually reinforce shifts in approaches to curriculum design, pedagogy and assessment.

Realising a broad interpretation of STEM education will also include aspects such as:

- working to develop a shared language around the purposes and practices of STEM education;
- identifying a shared framework for STEM education, including its goals and practices, that can accommodate diverse interests and needs of learners and schools;
- recognising the crucial role of teachers as professionals, and supporting teachers to innovate, develop and trial new curriculum and practices;
- supporting teacher capacities and interest around cross-disciplinary and cross professional (i.e., those outside of schools) learning in STEM;
- identifying school leadership practices that create conditions for teacher experimentation, learning and sustainable change for STEM education; and

Consequently, there is a need to create opportunities and conditions for appropriate and sustained teacher learning that enables and promotes change. Such teacher learning in STEM education will need to examine effective approaches to STEM education, potential outcomes, and possible future directions.

11.6 Concluding Thoughts

STEM as a world-wide movement needs to attend to the nature and purposes of 'STEM as STEM education' that align with, and help to clarify, the purposes of schooling. As Tan (Chap. 10) points out, this would mean embracing a "humanistic vision of education, broadly based on the notion that education should strive to give to students the greatest latitude for action in the future". This book is implicitly predicated on the premise that we must reconsider goals for schooling, and STEM in particular, but not in isolation. In this book, we have included the need to see cross-curriculum capabilities as central and fundamental goals, not as an add-on, if 'STEM' is to be more than a relabeling of S, T, E and M. Specifically, cross-curriculum capabilities need to be seen as central and fundamental goals of formal education. Integrated approaches to STEM education, when carefully planned and effectively implemented, can support multiple learning outcomes.

References

Berry, A., Gunstone, R., Loughran, J., & Mulhall, P. (2001). Using laboratory work for purposeful learning about the practice of science. In H. Behrendt, H. Dahncke, R. Duit, W. Gräber, M. Komorek, A. Kross, & P. Reiska (Eds.), *Research in science education – Past, present, and future* (pp. 313–318). Springer. https://doi.org/10.1007/0-306-47639-8_44

Biesta, G. J. J. (2013). *The beautiful risk of education*. Taylor & Francis.

Blackley, S., & Howell, J. (2015). A STEM narrative: 15 years in the making. *Australian Journal of Teacher Education, 3*, 102–111. https://doi.org/10.14221/ajte.2015v40n7.8

Corrigan, D. (2020). *Spotlight report 2: Implementing and integrated STEM education in schools – Five key questions answered*. Monash University. Retrieved from https://www.monash.edu/__data/assets/pdf_file/0007/2479444/sr2-stem-education-education-corrigan.pdf

Fensham, P. (2011). Knowledge to deal with the challenges to science education from without and within. In D. Corrigan, J. Dillon, & R. Gunstone (Eds.), *The professional knowledge base of science teaching* (pp. 295–318). Springer.

Lehrer, R. (2016, October). Perspectives on integrating elementary STEM education. Paper delivered at the forum: *Putting STEM education under the microscope*. Deakin University, Melbourne. Retrieved from https://blogs.deakin.edu.au/steme/wp-content/uploads/sites/39/2017/04/Lehrer-Deakin_STEM_2_reduced.pdf

Marginson, S., Tytler, R., Freeman, B., & Roberts, K. (2013). *STEM: Country comparisons. International comparisons of science, technology, engineering and mathematics (STEM) education* (Final report). Australian Council of Learned Academies.

Organisation for Economic Development (OECD). (2005). *The definition and selection of key competencies: Executive summary*. OECD. Retrieved from http://www.oecd.org/dataoecd/47/61/35070367.pdf

Rennie, L. J. (1990). Gender and science and technology in primary and secondary schooling: A personal summary of frame 1. In I. Granstam & I. Frostfeldt (Eds.), *Report book of the European and third world gender and science and technology conference* (pp. 190–194). Jönköping.

Rennie, L. (2020). Communicating certainty and uncertainty in science in out-of-school contexts. In D. Corrigan, C. Buntting, A. Fitzgerald, & A. Jones (Eds.), *Values in science education: The shifting sands* (pp. 7–30). Springer.

Rennie, L., Venville, G., & Wallace, J. (2012). Reflecting on curriculum integration: Seeking balance and connection though a worldly perspective. In L. Rennie, G. Venville, & J. Wallace (Eds.), *Integrating science, technology, engineering, and mathematics: Issues, reflections, and ways forward* (pp. 123–142). Routledge.

Ritchie, S. M. (2019). STEM education. In *Oxford research encyclopedia of education*. https://doi.org/10.1093/acrefore/9780190264093.013.237

Sahlberg, P. (2011). The fourth way of Finland. *Journal of Educational Change, 12*, 173–185. https://doi.org/10.1007/s10833-011-9157-y

Siegel, D., & Giamellaro, M. (2019). Defining STEM within a school district: A co-constructed and evolving process. *Cultural Studies of Science Education, 15*(1), 739–773. https://doi.org/10.1007/s11422-019-09959-2

Smith, D. V., & Gunstone, R. F. (2009). Science curriculum and the market liberal society of the 21st century: 'Re-visioning' the idea of science for all. *Research in Science Education, 39*, 1–16.

Tyler, R. W. (1949). *Basic principles of curriculum and instruction*. [Twenty-ninth impression, 1969]. The University of Chicago Press.

Zeidler, D. L., Herman, B. C., Clough, M. P., Olson, J. K., Kahn, S., & Newton, M. (2016). Humanitas emptor: Reconsidering recent trends and policy in science teacher education. *Journal of Science Teacher Education, 25*, 465–476. https://doi.org/10.1007/s10972-016-9481-4

The manufacturer's authorised representative in the EU is Springer Nature Customer Service Centre GmbH, Europaplatz 3, 69115 Heidelberg, Germany. If you have any concerns regarding our products, please contact ProductSafety@springernature.com

Printed and bound by CPI Group (UK) Ltd, Croydon, CR0 4YY
30/09/2026
02225471-0015